Advances in Photonic Materials and Devices

Technical Resources

Journal of the American Ceramic Society

www.ceramicjournal.org

With the highest impact factor of any ceramics-specific journal, the *Journal of the American Ceramic Society* is the world's leading source of published research in ceramics and related materials sciences.

Contents include ceramic processing science; electric and dielectic properties; mechanical, thermal and chemical properties; microstructure and phase equilibria; and much more.

Journal of the American Ceramic Society is abstracted/indexed in Chemical Abstracts, Ceramic Abstracts, Cambridge Scientific, ISI's Web of Science, Science Citation Index, Chemistry Citation Index, Materials Science Citation Index, Reaction Citation Index, Current Contents/ Physical, Chemical and Earth Sciences, Current Contents/Engineering, Computing and Technology, plus more.

View abstracts of all content from 1997 through the current issue at no charge at www.ceramicjournal.org. Subscribers receive full-text access to online content.

Published monthly in print and online. Annual subscription runs from January through December. ISSN 0002-7820

International Journal of Applied Ceramic Technology

www.ceramics.org/act

Launched in January 2004, *International Journal of Applied Ceramic Technology* is a must read for engineers, scientists,and companies using or exploring the use of engineered ceramics in product and commercial applications.

Led by an editorial board of experts from industry, government and universities, *International Journal of Applied Ceramic Technology* is a peer-reviewed publication that provides the latest information on fuel cells, nanotechnology, ceramic armor, thermal and environmental barrier coatings, functional materials, ceramic matrix composites, biomaterials, and other cutting-edge topics.

Go to www.ceramics.org/act to see the current issue's table of contents listing state-of-the-art coverage of important topics by internationally recognized leaders.

Published quarterly. Annual subscription runs from January through December.
ISSN 1546-542X

American Ceramic Society Bulletin

www.ceramicbulletin.org

The *American Ceramic Society Bulletin*, is a must-read publication devoted to current and emerging developments in materials, manufacturing processes, instrumentation, equipment, and systems impacting the global ceramics and glass industries.

The *Bulletin* is written primarily for key specifiers of products and services: researchers, engineers, other technical personnel and corporate managers involved in the research, development and manufacture of ceramic and glass products. Membership in The American Ceramic Society includes a subscription to the *Bulletin*, including online access.

Published monthly in print and online, the December issue includes the annual *ceramicSOURCE* company directory and buyer's guide. ISSN 0002-7812

Ceramic Engineering and Science Proceedings (CESP)

www.ceramics.org/cesp

Practical and effective solutions for manufacturing and processing issues are offered by industry experts. CESP includes five issues per year: Glass Problems, Whitewares & Materials, Advanced Ceramics and Composites, Porcelain Enamel. Annual subscription runs from January to December.
ISSN 0196-6219

ACerS-NIST Phase Equilibria Diagrams CD-ROM Database Version 3.0

www.ceramics.org/phasecd

The ACerS-NIST Phase Equilibria Diagrams CD-ROM Database Version 3.0 contains more than 19,000 diagrams previously published in 20 phase volumes produced as part of the ACerS-NIST Phase Equilibria Diagrams Program: Volumes I through XIII; Annuals 91, 92 and 93; High Tc Superconductors I & II; Zirconium & Zirconia Systems; and Electronic Ceramics I. The CD-ROM includes full commentaries and interactive capabilities.

Advances in Photonic Materials and Devices

Ceramic Transactions Volume 163

Proceedings of the 106th Annual Meeting of The American Ceramic Society, Indianapolis, Indiana, USA (2004)

Editor
Suhas Bhandarkar

Published by
The American Ceramic Society
PO Box 6136
Westerville, Ohio 43086-6136
www.ceramics.org

Advances in Photonic Materials and Devices

For information on ordering titles published by The American Ceramic Society, or to request a publications catalog, please call 614-794-5890, or visit our website at www.ceramics.org

ISBN 1-57498-184-6

Contents

Preface

As with other device technologies, photonics represents the confluence of materials, optical and device engineering and hence a symposium in photonics is most effective when it brings forth free exchange of thoughts from each of these disciplines even as only one of the fields serves as the host of the symposium.

It is in this spirit that the symposium on Photonics Materials and Devices was organized at the 106th Annual Meeting of The American Ceramic Society, April 18-21, 2004 in Indianapolis, Indiana. The symposium organizers were Dr. Suhas Bhandarkar from Alfred University in Alfred, NY, Dr. Glen Kowach from CCNY in NY, Dr. Ishwar Agarwal from NRL in Washington, D.C., and Dr. Burtrand Lee from Clemson University. There were five sessions in the symposium, listed here to provide the reader a sense of the subjects presented:

- Advances in optical waveguide materials and technology
- Non-linear optical materials and devices
- Novel photonic materials
- Photonic crystals and related applications
- Light Emitting Materials

A total of 34, both invited and submitted, papers were presented. Unfortunately, the uncertain financial and security climate that prevailed over the past year curtailed the papers from researchers abroad. A total of 10 papers are published in this volume. These papers are all peer-reviewed and represent a good cross section of the various subjects discussed in the symposium.

The remarkable advances made in photonics continue to grow despite a depressed telecommunications marketplace, that traditionally has been the major recipient of this technology. In particular, this Ceramic Transactions volume showcases the transformation of the photonics from a telecom-aligned technology to seeking a much wider sphere of applications.

Suhas Bhandarkar

NOVEL OXIDE GLASS AND GLASS CERAMIC MATERIALS FOR OPTICAL AMPLIFIER

Setsuhisa Tanabe
Graduate school of Human and Environmental Studies,
Kyoto University, Sakyo-ku, Kyoto 606-8501, Japan
Email: stanabe@gls.mbox.media.kyoto-u.ac.jp

ABSTRACT
Recent research results on novel oxide glass and glass ceramics materials for optical amplifiers are reviewed. In a series of Er-doped glasses with broad 1.5μm emission, heavy-metal oxide glasses are attracting great interest. Anomalous compositional dependences of optical properties of Er-doped antimony silicate glasses are shown. Structural implication of small compositional dependence observed for the Judd-Ofelt Ω_6 parameter and local phonon energy are discussed. Possibilities of U-band (1.65μm) amplifier in Er-doped oxide glass ceramics are also discussed. Finally, potential of Cr^{4+}-doped materials for ultra-broad amplifier is indicated in transparent glass ceramics, where the quantum efficiency of the 3d-emission becomes much improved compared with that of as-made glass.

1. INTRODUCTION

Due to rapid increase of information traffic and the need for flexible networks, there exists urgent demand for optical amplifiers with a wide and flat gain spectrum in the telecommunication window, to be used in the wavelength-division-multiplexing (WDM) network system. After the invention of the Er-doped fiber amplifier (EDFA), various types of amplifier devices have been developed in order to broaden the telecommunication bandwidth in the WDM network [1]. Tm^{3+}-doped and Pr^{3+}-doped fluoride fiber amplifiers have been developed for the S-band and O-band applications, respectively. Long-term reliability of fluoride fibers is still an issue for practical use [2]. The Raman amplifier composed of conventional silica fiber is also becoming practical use in WDM system that requires small gain (~10dB) in broad wavelength range. The gain range and bandwidth can be controlled by the wavelength and configuration of pumps. However, the pump power required is very high compared with the rare-earth doped fiber amplifiers. Still the rare earth doped amplifiers can be promising in the practical system due to their high power conversion efficiency.

Most of the EDFA utilized at present is made of silica-based glass fiber, where doped Er^{3+} ions show narrow emission band at 1.55μm resulting in narrow gain spectra. Following the report of wide spectra [3], Mori reported excellent performance of a tellurite based EDFA [4], which shows 80nm-wide gain around 1.53 ~ 1.61μm. In order to increase the channels and to improve the performance of WDM network, it is important to investigate a material with wider gain spectra. We have reported that the Bi_2O_3-based borosilicate glass showed broad emission spectra of the 1.55μm transition and large Ω_6 of Er^{3+} ion [5]. In 2001, a group of Corning reported the MCS (multi-component silicate) glass containing Sb_2O_3, which shows wider gain than the glasses [6]. The important factor dominating the cross section and its bandwidth is the Judd-Ofelt Ω_6 parameter of Er^{3+} ions [7] as well as the refractive index. We also reported a strong correlation between the Ω_6 and the ionicity of Er-ligand bond in various glasses and its origin [8.9]. Therefore, it is interesting to investigate glass systems giving ionic ligand fields, which would attain a large Ω_6.

In the first part of this paper, we report the optical properties of the oxide glass compositions based on Sb_2O_3, in which Er^{3+} ions show very broad emission, and the local structure of rare-earth ion in this glass.

In the second part, possibility of U-band amplifier based on oxide glass ceramics is discussed and an example of Er:YAG glass ceramics is presented. The origin of energy level splitting of $^{2S+1}L_J$ multiplets of rare earth ions should be considered for spectral design.

It is well understood that the typical gain bandwidth of the rare earth doped amplifiers is usually much less than 100nm in near-infrared region, largely due to the nature of 4f energy levels. On the other hand, the 3d-transition metal ions show broadband emission depending on the host materials. The Cr^{4+} ion, an anomalous valence state of chromium, is also a broadband luminescent center in solid-state materials. Since its emission range, typically around 1.2~1.6 μm, covers the optical telecommunication window (Fig.1), Cr^{4+}-activated materials can be a fair candidate for novel broadband amplifier.

Glass hosts are favorable in its capability of fiber fabrication, while luminescence efficiencies of Cr^{4+} in them are generally much lower than in crystalline hosts [10]. In this regard, transparent glass-ceramics with Cr^{4+}-doped crystals would be desirable host materials.

In the third part, we report a novel transparent Cr^{4+}-doped glass-ceramics developed recently [11]. The precipitated crystal is gehlenite ($Ca_2Al_2SiO_7$), in which Cr^{4+} can be stabilized and show a broadband luminescence around 1.1~1.4 μm, centered at 1.24 μm. The spectroscopic investigation of the novel glass ceramics are shown.

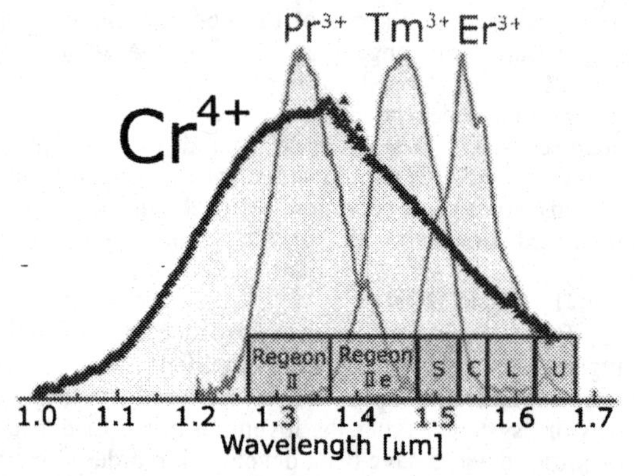

Fig.1. Comparison of emission spectra of Cr^{4+} and rare earth ions in tellurite glass [26].

2. Er-DOPED Sb_2O_3-GLASSES

2.1. Preparation and Characterization

Glasses in the composition of xSb_2O_3-$3Al_2O_3$-$(97-x)SiO_2$ (x=27,37,47,57,67) doped with 0.5mol% Er_2O_3 and those doped with 0.5mol% Eu_2O_3 were prepared by melting mixed powders in an alumina crucible at 1100-1450°C. The obtained glasses were cut into 15x10x3 mm^3 shape before polishing into optical surface.

Emission spectra were measured by using a 970nm laser diode (LD) and an InGaAs photodiode (160kHz). The lifetime of the 1.53μm emission was measured by pumping with light pulses of the LD and the time evolution of the signal of the detector was collected with a 100 MHz digital oscilloscope. Decay curves were analyzed by a least-square fitting to get the lifetime.

The number of Er^{3+} ions in unit volume, ρ_N was calculated with the molecular weight and density. Density of the obtained glass was measured by the Archimedes method using kerosene as an immersion liquid. The refractive index, $n(\lambda)$ was measured by a prism coupling method at wavelength of 633nm, 1304nm and 1550nm. Absorption spectra were measured in 400nm ~ 1700nm with Shimadzu UV-3101PC spectrophotometer. With an integrated area of the absorption band, spontaneous emission probability, A of the 1.5μm was calculated by,

$$A(J \rightarrow J') = \frac{(2J+1)8\pi c n^2}{(2J'+1)\bar{\lambda}^4} \times \int_{.4\mu m}^{1.7\mu m} \frac{k(\lambda)}{\rho_N} d\lambda \qquad (1)$$

where c is the light velocity, $\bar{\lambda}$ is the mean wavelength of emission, J and J' are the total momentum for the upper and lower levels, $k(\lambda)$ is the absorption coefficient, and n was the refractive index at wavelength of 1530nm.

The Judd-Ofelt parameters of Er^{3+} ions were calculated by the method described elsewhere [8] with cross sections of five intense bands ($^4F_{7/2}$, $^2H_{11/2}$+ $^4S_{3/2}$, $^4F_{9/2}$, $^4I_{11/2}$, $^4I_{13/2}$) in 470nm ~ 1700nm.

The excitation spectra of the $^5D_0 \rightarrow ^7F_2$ emission at 613nm of Eu^{3+} doped glasses was measured in the range of 440 ~ 470 nm. The phonon sideband associated with the pure electronic the $^5D_2 \leftarrow ^7F_0$ transition 465nm was multiplied by 50 times to investigate the phonon mode coupled to rare-earth ions [12], which contributes to multiphonon relaxation.

2.2. Properties of Glass and Spectroscopy of Rare Earth Ions

Fig.2 shows the compositional dependence of refractive index, $n(\lambda)$ of the glasses at 633nm and 1550nm increased with decreasing wavelength. The n of the xSb_2O_3-$3Al_2O_3$-(97-x) SiO_2 glasses (in mol%) at 1.55μm was 1.66 ~ 1.90, which increased with increasing Sb_2O_3 content.

Fig.3 shows emission spectra of the $^4I_{13/2} \rightarrow ^4I_{15/2}$ of Er^{3+} ions in the xSb_2O_3-$3Al_2O_3$-(97-x) SiO_2 glasses, in the $75TeO_2$-$20ZnO$-$5Na_2O$-$0.5Er_2O_3$ and in the $95SiO_2$-$5GeO_2$-$0.5Er_2O_3$ as a reference. The spectrum of the silica shows the narrowest bandwidth of about 30nm and that of the tellurite was 60nm width. It is seen that the Sb_2O_3-based glasses show spectra comparable or broader than that of the tellurite.

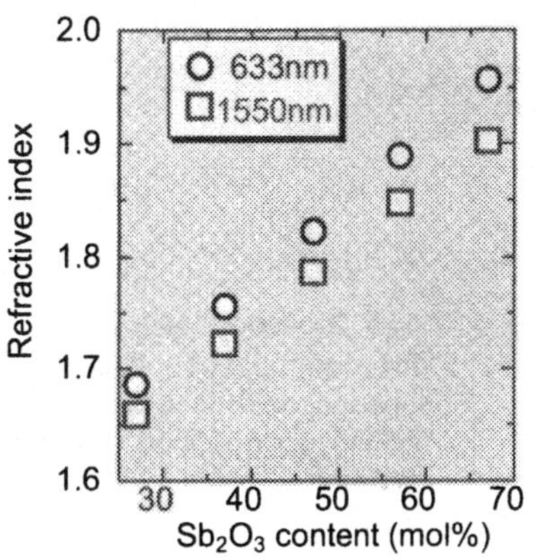

Fig.2. Compositional dependence of refractive index in xSb_2O_3•(97-x) SiO_2•$3Al_2O_3$•$0.5Er_2O_3$ glasses

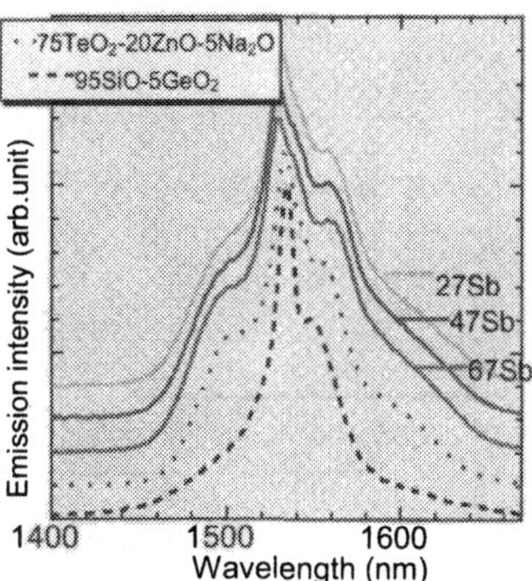

Fig.3. Fluorescence spectra of xSb_2O_3•(97-x) SiO_2•$3Al_2O_3$•$0.5Er_2O_3$ glasses

Fig.4 shows the compositional dependence of fluorescence lifetime, τ_f of the $^4I_{13/2}$ level in the xSb_2O_3-$3Al_2O_3$-$(97-x)$ SiO_2-0.5 Er_2O_3 glasses. In the range of Sb_2O_3 content x=37-67, the fluorescence lifetimes of the $^4I_{13/2}$ level were almost unchanged, and about 2.5ms, whereas in the Sb_2O_3 content x=27, the lifetime was dramatically decreased.

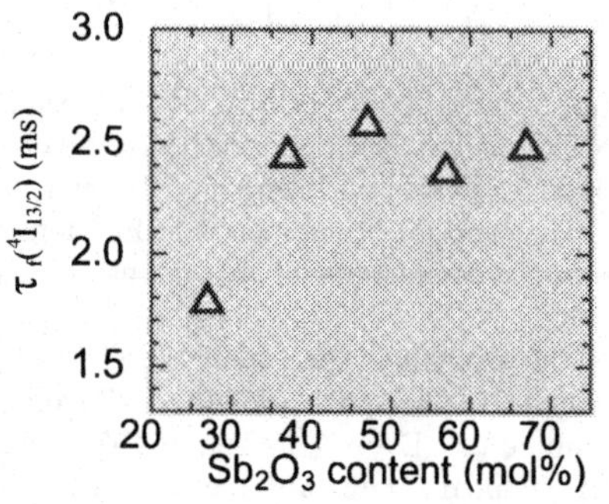

Fig.4. Compositional dependence of lifetime of Er^{3+}: $^4I_{13/2}$ level in glasses.

The excitation spectrum associated with Eu^{3+}: $^5D_2 \leftarrow {}^7F_0$ transition for the xSb_2O_3-$3Al_2O_3$-$(97-x)$ SiO_2-0.5 Er_2O_3 glasses are shown in Fig.5. The intense band due to the pure electronic transition (PET) Eu^{3+}: $^5D_2 \leftarrow {}^7F_0$ transition is located around 464nm, while the phonon sideband (PSB) coupled to the rare earth ions is observed in the higher-energy range [8]. The position and shape of the phonon sideband were almost unchanged with Sb_2O_3 content.

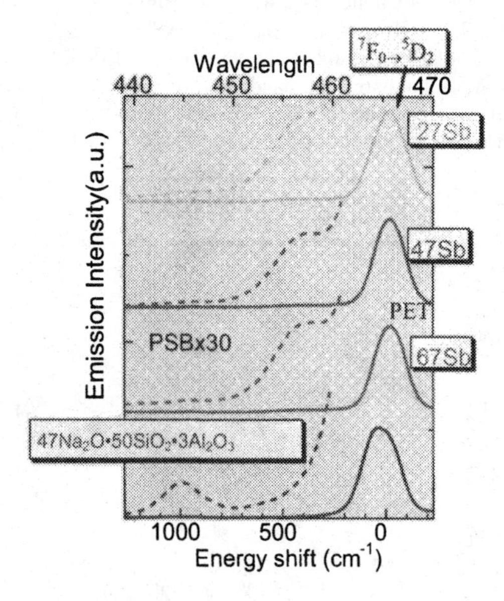

Fig.5. Excitation spectra of Eu^{3+}-doped glasses. Phonon sideband can be seen in the higher-energy- side of $^5D_2 \leftarrow {}^7F_0$ transition.

2.3. Local Structure of Rare Earth Ions in Sb_2O_3-Glass

Fig.6 shows the calculated spontaneous emission probability, A of the $^4I_{13/2} \rightarrow {}^4I_{15/2}$ in the xSb_2O_3-$3Al_2O_3$-$(97-x)$ SiO_2-0.5 Er_2O_3 glasses. The A also increased with increasing x, being nearly 200 s^{-1}. The large A is mainly due to large n of the host glasses, which are composed of large amount of Sb^{3+} ion, a p-block element of $5s^2$ electrons having large polarizability [13].

With the measured lifetime and $A_{JJ'}$ from Eq.(1), radiative quantum efficiency, η, were calculated by

$$\eta = \frac{\sum A}{\sum A + \sum W_{NR}} = \tau_f \sum A \qquad (2)$$

and plotted in Fig.7. Reflecting the compositional variations of $A_{JJ'}$, the η increased with increasing Sb_2O_3 content, x. The η values are relatively small compared with those of EDFA's ever reported, but still much larger than that of Pr-doped fiber amplifiers (4%), which perform large gain at 1.3μm[14]. These low values obtained may be due to lower estimation of real local refractive index, i.e., deviation from measured average index. This can be related with no compositional dependence of the local phonon energy obtained from phonon sideband spectra.

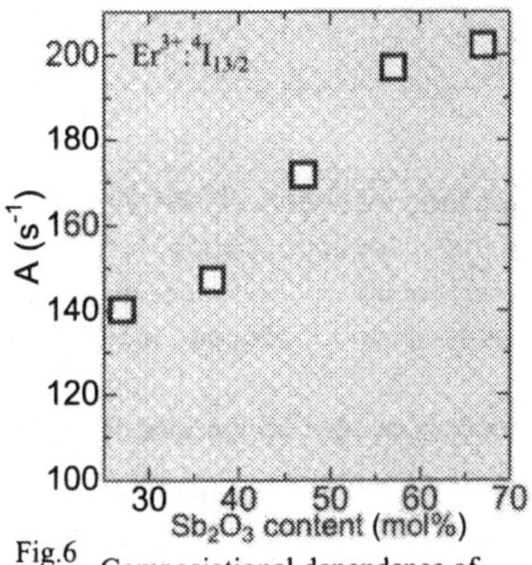

Fig.6 Composiotional dependence of spontaneous emission rate of Er^{3+}:$^4I_{13/2}$ in xSb_2O_3-$(97-x)SiO_2$-$3Al_2O_3$-$0.5Er_2O_3$ glasses.

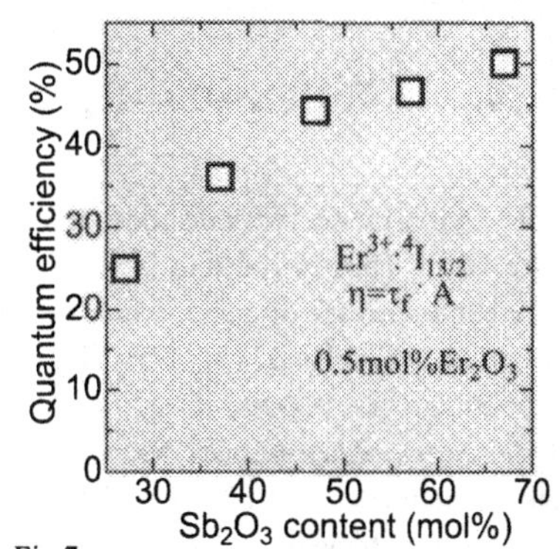

Fig.7 Composiotional dependence of quantum efficiency of Er^{3+}:$^4I_{13/2}$ in xSb_2O_3-$(97-x)SiO_2$-$3Al_2O_3$-$0.5Er_2O_3$ glasses.

Fig.8 shows that in the Judd-Ofelt treatment obtained by using the five electric-dipole bands. It is seen that the Ω_6 values were almost unchanged with Sb_2O_3 content. These results suggest that the Er^{3+} ions are surrounded selectively by Sb_2O_3-rich phase. Generally the A of the $^4I_{13/2}$-$^4I_{15/2}$ band is related with the line strengths, S of electric-dipole (ED) and magnetic-dipole (MD) components by[15],

$$A_{JJ'} = \frac{64\pi^4 e^2}{3h(2J'+1)\lambda^3}\left\{\frac{n(n^2+2)^2}{9} \times S_{JJ'}^{ED} + n^3 \times S_{JJ'}^{MD}\right\} \qquad (3)$$

where e is the elementary charge, h is the Planck constant. The MD transition is independent of the ligand field and contributes to a sharp central peak of spectra around 1.53μm. Because the S^{MD} is characteristic only to the transition determined by the quantum numbers [16], one of the important factors affecting the compositional variations of the emission properties is the ED transition. The S^{ED} of the $^4I_{13/2}$-$^4I_{15/2}$ is obtained with the Judd-Ofelt parameters and reduced matrix elements by [17,18],

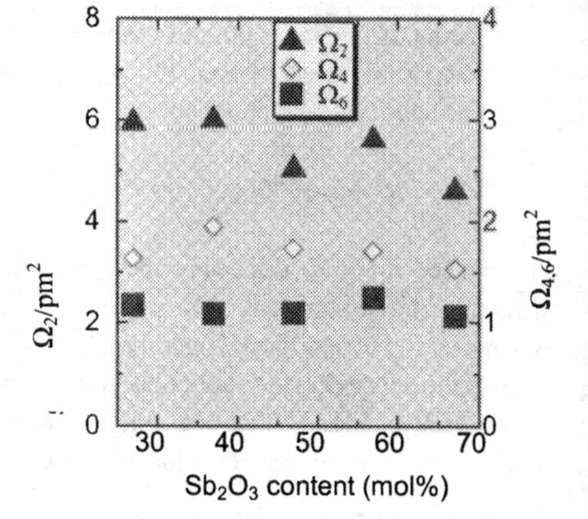

Fig.8. Composiotional dependence of Judd-Ofelt Ω_t parameters of Er^{3+} ions

$$S^{ed}[^4I_{13/2};^4I_{15/2}]=0.019\Omega_2+0.118\Omega_4+1.462\Omega_6 \quad (4)$$

According to Eq.(4), the Ω_6 plays the most dominant role on the cross section of the 1.5μm band among three Ω_t's. Thus in order to increase the bandwidth of spectra, which is varied with local structure, the increase of the Ω_6 would be effective, because the ED contributes to broader component of the 1.5μm band [18]. The large Ω_6 value and refractive index may contribute to broad bandwidth in these glasses.

Fig.9 shown the compositional dependence of the phonon energy, hω obtained from the wavelength of the phonon sideband and that of the pure electronic transition. The phonon energy was found to be about 400cm^{-1} in all the compositions up to 70mol% SiO_2 content. Usually in most silicate glasses, the Si-O stretching mode is coupled to the rare earth ions even in low SiO_2 composition. The present results are in contrast to the above facts and thus suggest that the Er^{3+} ions are surrounded selectively by Sb_2O_3-rich phase and are not affected by Si-O .

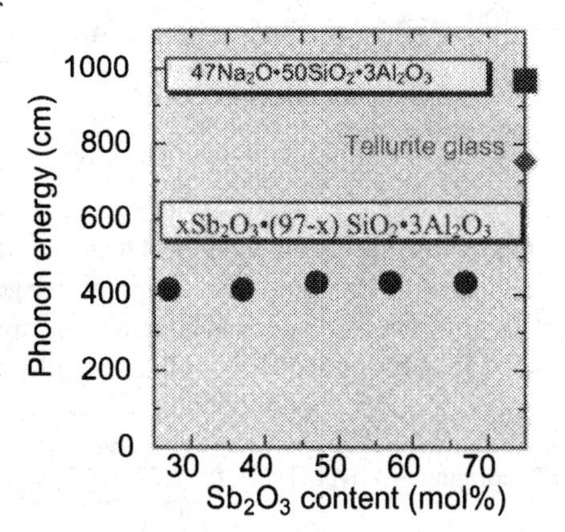

Fig.9. Composiotional dependence of vibration with about 1000cm^{-1} energy phonon energy from PSB spectra

2.4. Origin of "Non-Silicate" Environment

Fig.10 show the TEM image and a structural model of the glass, where the nano-scale phase separation can be observed [19]. The dark region can be Sb_2O_3-rich phase and the other can be SiO_2-rich phase. Since the solubility of rare earth ions in a pure silica or silica-rich glass is very low [20], Er^{3+} ions can be condensed in the Sb_2O_3-rich phase, as indicated in the spectroscopic results mentioned above. The concentration dependence of lifetime of Er^{3+}:$^4I_{13/2}$ shows more rapid decrease in a Sb_2O_3-poor composition, indicating that Er^{3+} ions are more condensed. The tendency is moderate in glasses with Sb_2O_3-rich compositions.

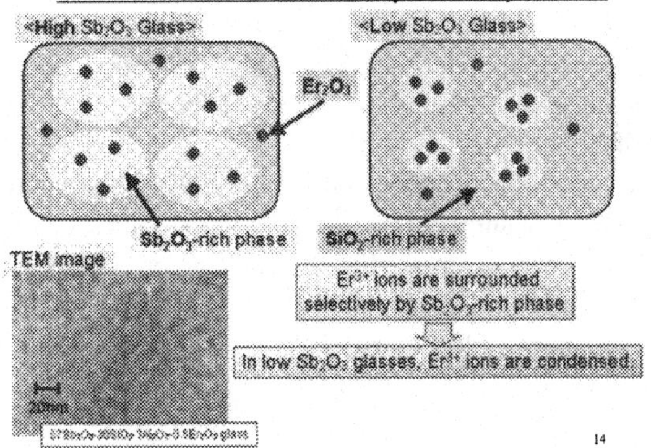

Fig.10. TEM image of the glass and structural model of nano-scale phase separation [19].

3. POTENTIAL MATERIALS FOR U-BAND AMPLIFIER

3.1. Rare Earth Candidates For 1.6μm Emission

Two rare earth ions have been reported to show emission in U-band wavelength range. Choi reported 1.6μm emission for the Pr^{3+}:$^4F_{3,4} \rightarrow {}^3H_4$ in a selenide glass [21] and Lee reported the Ho^{3+}:$^5I_5 \rightarrow {}^5I_7$ transition in a sulfide glass [22]. Fig.10 shows the energy level diagrams of candidate ions with 1.6μm transition [23]. As can be seen from Fig.11, the energy gap of the Pr Pr^{3+}:$^4F_{3,4}$ and Ho^{3+}:$^5I_5$ are 1400cm-1 and 2500cm-1, respectively. The gain efficiency of amplifiers is dominated by the quantum efficiency of the initial level, which is largely dependent on the multiphonon decay loss. For the levels with the energy gap to the next lower level, ΔE is 3000cm^{-1}, oxide glasses with phonon energy, ħω, higher than 600cm-1 cannot be a practical host with good efficiency, because the multiphonon decay rate, W_p increases drastically when ΔE/ħω is less than 5. In the case of U-band emissions for Pr^{3+} and Ho^{3+}, even the phonon energy of typical fluoride glasses such as ZBLAN (~500cm^{-1}) is too high to suppress the nonradiative loss. That is the main reason why the 1.6μm emission is reported only in chalcogenide glasses, ħω of which is less than 400cm^{-1}. However, the fiberizability and reliability of nonoxide glasses are issue for practical application. On the other hand, ΔE of the Er^{3+}:$^4I_{13/2}$ is 6500cm^{-1}, large enough to obtain quantum efficiency over 90% even in oxide hosts with high phonon energy.

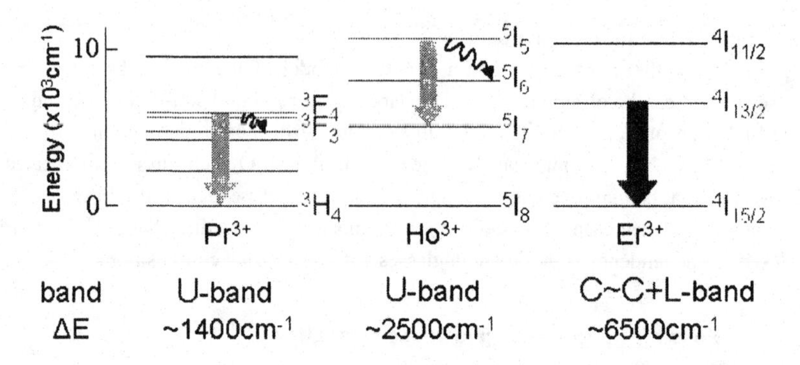

Fig.11 . Energy level diagrams of Pr^{3+}, Ho^{3+}, and Er^{3+} ions. The energy gap of the initial level of 1.6μm emission is also shown.

3.2. Do We Really Know Spectra of Er^{3+}?

In addition to the C-band and L-band EDFA, the S-band EDFA was reported in a conventional silica-based EDF [24]. It requires a special and tricky pumping configuration with several C-band ASE-suppression filters and much higher pumping power is required than normal EDFAs for the C+L band. Due to the energy distribution, mainly to the Stark level structure of major Er^{3+} ions in glasses, most of glass-based EDFAs usually cover only C+L-band (1520~1610nm) by normal pumping scheme.

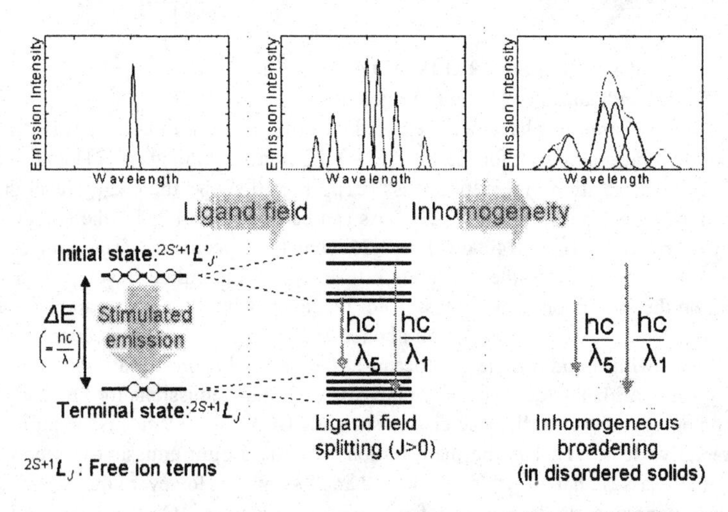

Fig.12 Origin of spectral broadening due to energy level splitting of $^{2S+1}L_J$ state of lanthanide ions in crystalline ligand field and in disorderd solids.

Fig.12 shows the origin of spectral broadening due to energy level splitting of $^{2S+1}L_J$ state of lanthanide ions in crystalline ligand field and in disorderd solids. In a crystalline host, depending on its structure, a unique ligand field can be expected for anomalous Stark splitting. Possibility of homogeneous doping of Er^{3+} ions in crystals is determined by the availability of crystallographic sites suitable for lanthanide substitution. Crystals composed of Y^{3+}, La^{3+} or Gd^{3+} ions, such as yttrium aluminum garnet, YAG, can accommodate substantial amount of other optically 4f-active lanthanide ions in the rare earth site. The spectral shape of doped crystals usually become discrete, while those of doped glasses are continuously broadened. The spectral features of glasses are advantageous for obtaining a flat gain spectrum as a WDM amplifier, but we might not be able to expect anomalous Stark splitting for Er^{3+} ions, since the absence of structural restriction of the crystallographic site in glass may usually result in well-observed "averaged" spectrum centered at 1530nm.

3.3. Line Broadening in Er:YAG Glass Ceramics

In 2003, we have reported the U-band emission in a glass ceramics containing Er:YAG [23]. Fig.13 shows the comparison of Er^{3+}: $^4I_{13/2}\rightarrow{}^4I_{15/2}$ emission in glasses and YAG crystal. Due to its unique Stark level structure, the emission bands are observed in the U-band range. Sharp spectral bands can be moderated by introducing inhomogeneity to the Y^{3+}-site in disordered structure by means of formation of solid solution. The $Ca_3Al_2(SiO_4)_3$ crystal has a cubic garnet structure, the lattice constant of which is only 1% different (a_0=11.849A) from that of cubic YAG (a_0=12.009A). Solid solution can be formed in the composition $Y_{3-x}Ca_xAl_{5-x}Si_xO_{12}$. Charge neutrality is maintained by substitution of same amount of Y^{3+} and Al^{3+} with Ca^{2+} and Si^{4+} at octahedral and tetrahedral sites, respectively. Er^{3+} ions can substitute the Y^{3+} site, the structural configuration of the second nearest cations around which can be varied in the solid solution. This variation results in variation of the ligand field based on the eight nearest oxygen ions and thus can lead to the inhomogeneous broadening of spectra. Our glass ceramics is a CaO-Y_2O_3-Al_2O_3-SiO_2 system, in which the Er:YAG phase contains a certain amount of Ca^{2+} and Si^{4+} [23]. As shown in Fig.14, the emission spectra of the glass ceramics are slightly broadened compared with that of Er:YAG crystal. More detailed studies are necessary to clarify the effect of nano-sized crystals and solid solution on the inhomogeneous broadening of emission spectra [25].

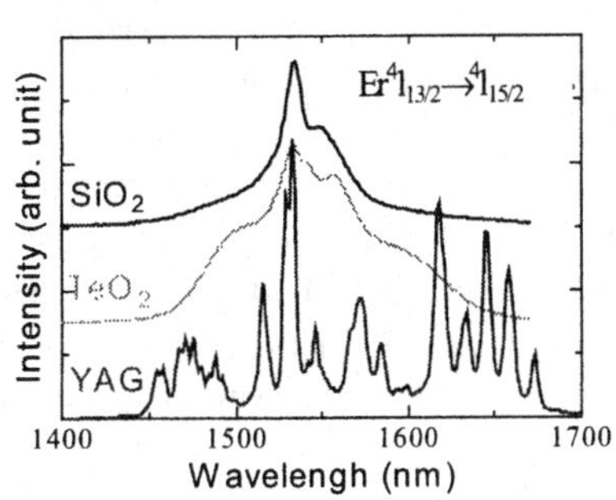

Fig.13. Emission spectra of Er^{3+} in glasses and YAG crystal.

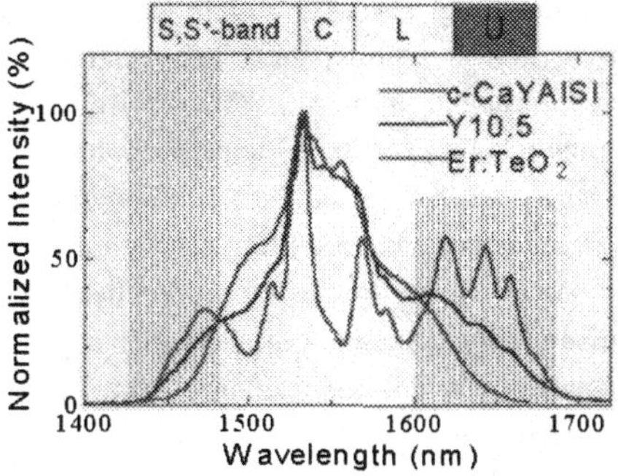

Fig.14. Emission spectra of Er:YAG glass ceramics and a tellurite glass.

4. Cr^{4+}-DOPED GLASS CERAMICS FOR ULTRA-BROAD AMPLIFIER

4.1. Why We Work on Cr^{4+} and Glass Ceramics as a Host

It is well understood that the typical gain bandwidth of the rare earth doped amplifiers is usually much less than 100nm in the near-infrared, largely due to the nature of 4f energy levels. On the other hand, the 3d-transition metal ions show broadband emission depending on the host materials. Some solid-state lasers such as Alexandrite (Cr^{3+}) and Ti-sapphire (Ti^{3+}) are known to be widely tunable in the near-infrared region. The Cr^{4+} ion, an anomalous valence state of chromium, is also a broadband luminescent center in solid-state materials [10]. Since its emission range, typically around 1.1~1.6 μm, covers the optical telecommunication windows in the future WDM, Cr^{4+}-activated materials can be a candidate for novel broadband amplifier [26].

However, host materials in which Cr^{4+} can be stabilized are limited. Until now, Cr^{4+}-activated tunable laser operations have been reported only in crystals such as Cr^{4+}:YAG [27] and Cr^{4+}:forsterite [28], and also broad-band luminescence of Cr^{4+} have been reported in calcium aluminosilicate glass [10,29].

Glass hosts are favorable in its capability of fiber fabrication [31], while luminescence efficiencies of Cr^{4+} in them are generally much lower than in crystalline hosts [32]. In this regard, transparent glass-ceramics with Cr^{4+}-doped crystals are expected to be desirable host materials. However, to our knowledge, there have been only a few studies to develop Cr^{4+}-doped glass ceramics; those are, Cr^{4+}: forsterite glass-ceramics [32], and Cr^{4+}: willemite glass- ceramics [33]. We reported a novel transparent Cr^{4+}-doped glass-ceramics developed. The precipitated crystal is gehlenite ($Ca_2Al_2SiO_7$), in which Cr^{4+} can be stabilized and show a broadband luminescence around 1.1~1.4 μm, centered at 1.24 μm. The preparing conditions and luminescence properties of this material have been investigated.

4.2. Sample Preparation and Characterization

As the starting material, a bulk glass in the composition of $50CaO-40Al_2O_3-10SiO_2-0.05Cr_2O_3$ (mol%) was prepared by conventional melting method. The batch was melted with a platinum crucible in air at 1873K for 1h. After annealed, the glass obtained was cut into proper size (about 10 x 10 x 3 mm), then heat-treated by two-step sequence; the first to cause homogeneous nucleation and the second to grow crystallites. Temperature and time conditions of heat treatment were varied for each samples, as shown in Table 1. These conditions were chosen based on the results of differential thermal analysis (DTA) of each sample. For comparison, polycrystalline samples of Cr-doped gehlenite ($Ca_2Al_2SiO_7$) and β-$CaAl_2O_4$ were also prepared by solid-state reaction. X-ray diffraction (XRD) of each sample was measured to identify the phases.

Absorption spectra were measured at room temperature, using a spectrophotometer. Emission spectra were also measured at room temperature, by using a 792 nm laser diode (LD) as a pumping source and an InGaAs photodiode as a detector in the range of 850 to 1670 nm.

The sensitivity calibration of this measurement system was done with a broad spectrum of a standard halogen lamp.

Table 1. Heat-treating schedules.

Sample name	Nucleation	Growth
A	--	--
B	1183K, 6.5h	1273K, 1h
C	1183K, 6.5h	1373K, 1h

4.3. Appearance of Glass Ceramics and XRD Pattern

Through heat treatment, all samples have remained mostly transparent but the colors have been obviously changed; while the sample A (as-quenched) is green, the sample B looks bluish green and the sample C shows deep blue. Fig. 15 shows the X-ray diffraction patterns of all samples. It is shown that two kinds of crystals, β-$CaAl_2O_4$ and gehlenite, have precipitated by heat treatment, preserving the transparency as a bulk. This result indicates that the temperature dependence of nucleation and growth rates of these two crystals may be quite similar. Actually, the DTA curve of each sample shows only one apparent exothermic peak. However, the color centers, namely the Cr ions, seem to be selectively incorporated in the gehlenite phase; when ground to a powder, the sample C exhibits light-blue just like the polycrystalline Cr: gehlenite, while the polycrystalline Cr:β-$CaAl_2O_4$ is yellow.

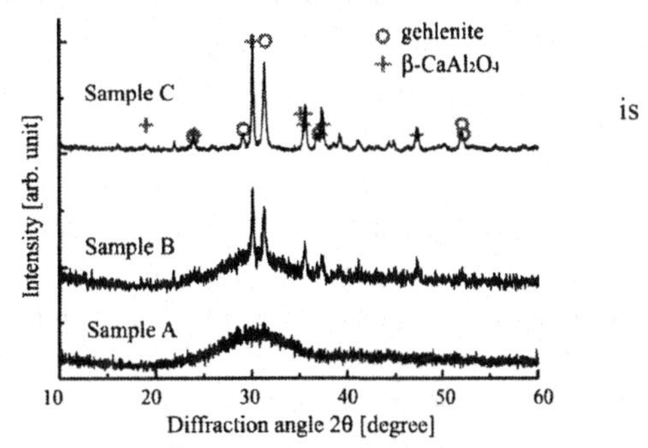

Fig. 15. XRD patterns of Cr doped glass and Cr:gehlenite glass-ceramics.

4.4. Absorption Spectra of Cr Ions

Absorption spectra of the samples A and C are shown in Fig. 16. The 370 nm band is attributed to the charge transfer transition of the Cr^{6+}[34]. As for the sample C, the 500~1200 nm bands can be ascribed to the Cr^{4+} ions incorporated in tetrahedral sites of randomly-directed gehlenite crystallites, referring to the polarized absorption spectrum of Cr^{4+}: gehlenite single crystal [35]. The sample A shows weaker bands in this region, which could be attributed to the

Cr^{4+} in tetrahedral sites in the glass [34], although the Cr^{3+} absorption bands may overlap with them as discussed later with the emission spectrum. Thus the incorporation of the Cr^{4+} in the gehlenite phase is indicated again, and it is expected that the Cr^{4+} concentration is considerably higher in the sample C than in the sample A.

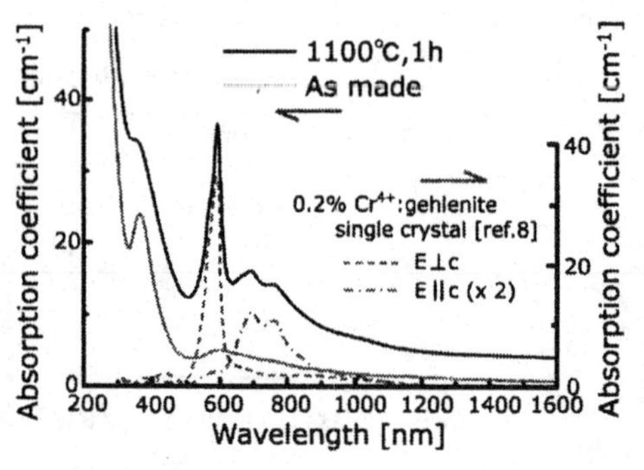

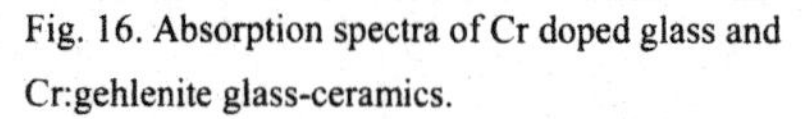
Fig. 16. Absorption spectra of Cr doped glass and Cr:gehlenite glass-ceramics.

4.5. Emission Spectra of Glass and Glass Ceramics

Emission spectra of three samples are shown in Fig. 17. The samples B and C show one emission band peaked at around 1240 nm, while the sample A shows two emission bands peaked at 1350 nm and 1000 nm. The emission intensity has obviously increased by heat treatment. The peak wavelength (λ_p) and full width at half maximum (w, FWHM) of the samples B and C are also shown in Fig. 17.

According to the previous studies [34], the tetrahedrally coordinated Cr^{4+} ions show the emission band at around 1350 nm with FWHM ~300 nm in calcium alumino-silicate glass, while the Cr^{3+} band is around 1000 nm in silicate glass. These are attributed to the transition from $^3T_2$ state to $^3A_2$ ground state of Cr^{4+} and that from $^4T_2$ to $^4A_2$ of Cr^{3+}, respectively. Hence, it is deduced that considerable amount of the Cr^{3+} ions coexist with the Cr^{4+} ions in the sample A, having its excitation band covering 792 nm (Fig. 18).

The polycrystalline Cr:gehlenite shows an emission band at around 1240 nm, whereas the polycrystalline Cr:β-$CaAl_2O_4$ exhibits an emission band at 1310 nm (Fig.19). Combined with the previous data for the Cr^{4+}:gehlenite single crystal [35], the 1240 nm emission band of the samples B and C is ascribed to the Cr^{4+} ions in gehlenite phase. Thus, it is confirmed that most of the Cr^{4+} and the Cr^{3+} ions in the as-quenched glass have been incorporated in the precipitated gehlenite crystallites as Cr^{4+}, through the heat treatment. Fig. 20 shows a simple diagram of this phenomenon.

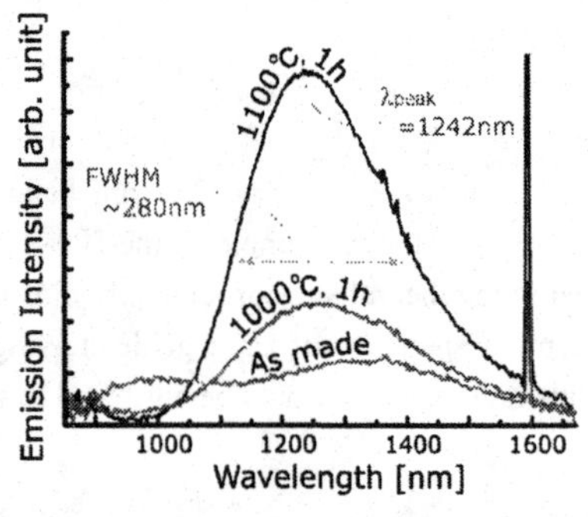

Fig. 17. Emission spectra of Cr doped glass and Cr:gehlenite glass-ceramics, excited at 792 nm.

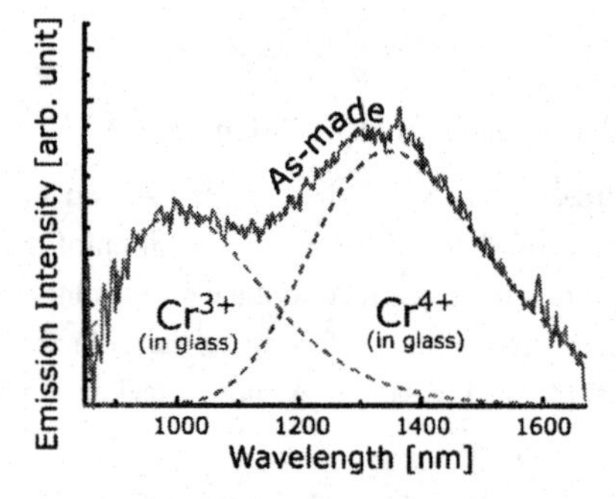

Fig. 18. Deconvolution of the emission spectrum of the sample A.

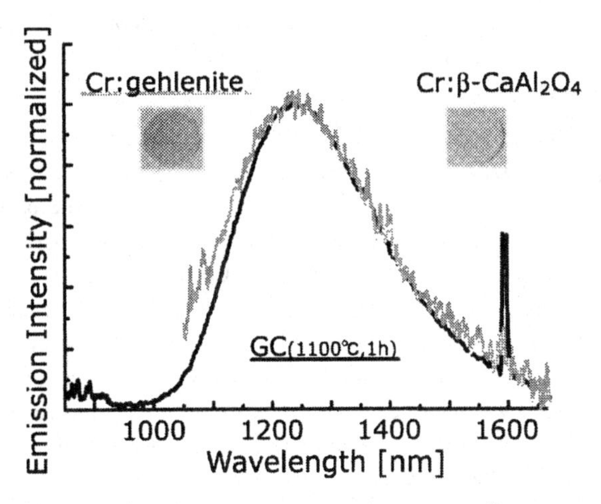

Fig.19. Comparison of emission spectra of Cr^{4+}-doped glass ceramics and polycrystalline samples.

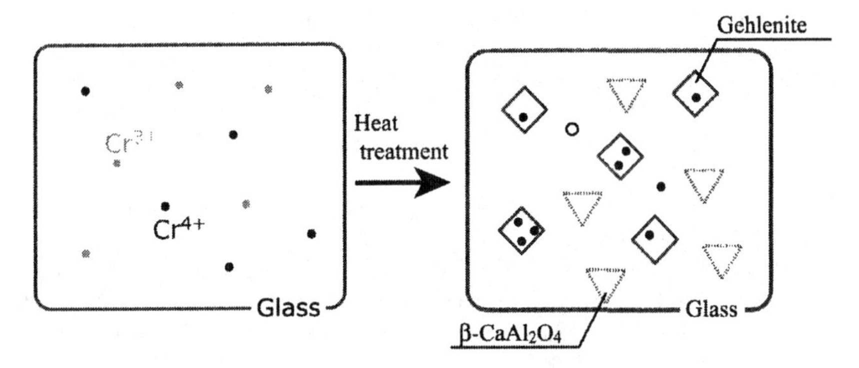

Fig. 20. Schematic model of states of Cr ions in glass and glass ceramics. While chromium ions are uniformly distributed as Cr^{3+} & Cr^{4+} in glass matrix, they become incorporated selectively in gehlenite crystallites as Cr^{4+}.

Obviously, there is a positive correlation between the emission intensity and the degree of crystallization, while the FWHM slightly decreases as the crystallization progresses (Figs. 15 and 17). This increase of emission intensity probably arises from the increase of the Cr^{4+} content and quantum efficiency in the crystalline ligand field. The greater FWHM of the sample B compared with the sample C may be a sign of the Cr^{4+} and the Cr^{3+} remaining in the amorphous phase as observed in XRD pattern of Fig. 15.

5. CONCLUSIONS

The Sb_2O_3-silicate glasses showed broad emission spectra of 1.5μm with a large cross section of Er^{3+} ions. Even the glasses with low Sb_2O_3 content showed broader emission band than the tellurite glass. It is suggested that the nano-scale phase separation exist in these glasses and the Er ions are selectively condensed in the Sb_2O_3-rich phase.

Transparent Cr^{4+}-doped gehlenite ($Ca_2Al_2SiO_7$) glass-ceramics was developed from a calcium aluminosilicate system by simple two-step heat treatment of as-melted glass. A broadband luminescence centered at 1240 nm with FWHM of ~300 nm was observed, with higher intensity than that in the as-made glass. There also precipitated β-$CaAl_2O_4$ in these samples, but it was indicated that Cr ions are selectively incorporated in the gehlenite phase as Cr^{4+}. The glass-ceramics containing crystallites of laser hosts can be a candidate for a novel broadband amplifier material.

Series of this work was carried out with great contribution of graduate students in Kyoto University, Mr. Masafumi Onishi, Masayuki Nishi, Hiroshi Yamazaki.

6. References

[1] S.Tanabe, "Rare Earth Doped Glasses for Fiber Amplifiers in Broad-band Telecommunication", *C.R.Chimie* 5, 815-824 (2002).
[2] S.Tanabe "Properties of Tm-doped tellurite glasses for 1.4μm amplifier", in *"Rare-Earth-Doped Materials and Devices V", SPIE* vol.4282, pp.85-92 (2001).
[3] J.S.Wang, E.M.Vogel, E.Snitzer, "Tellurite glass: a new candidate for fiber devices", *Opt. Mater.* **3**, 187 (1994).
[4] A.Mori, Y.Ohishi and S.Sudo, "Erbium-doped tellurite fiber laser and amplifier", *Electron. Lett.* **33**[10], 863-864 (1997).
[5] S. Tanabe, N. Sugimoto, S. Ito, T. Hanada, "Broad-band 1.5μm emission of Er^{3+} ions in bismuth-based oxide glasses for potential WDM amplifier", *J.Luminesc.* **87**, 670-672 (2000).
[6] J.Minelly, A.Ellison, "Applications of antimony-silicate glasses for fiber optic amplifiers", *Opt.Fiber Tech.* 8, 123-138 (2002).
[7] S.Tanabe, T.Hanada, "Local Structure and 1.5μm quantum efficiency of erbium doped glasses for optical amplifiers," *J.Non-Cryst. Solids*, **196**, 101-105 (1996).
[8] S.Tanabe, T.Ohyagi, T.Todoroki, T.Hanada, and N.Soga, "Relationship between the Ω_6 Intensity Parameters of Er^{3+} Ions and The ^{151}Eu Isomer Shift in Oxide Glasses", *J.Appl.Phys.* **73**[12] 8451-54 (1993).
[9] S.Tanabe, K.Takahara, M.Takahashi and Y.Kawamoto, "Spectroscopic studies on radiative transitions and upconversion characteristics of Er^{3+} ions in simple pseudo-ternary fluoride glasses MF_n-BaF_2-YF_3 (M: Zr, Hf, Al, Sc, Ga, In, or Zn)", *J. Opt. Soc. Am. B* **12**[5] 786-793 (1995).

[10] X.Wu, H.Yuan, W.M.Yen, B.G.Aitkin, "Compositional dependence of the luminescence from Cr^{4+}-doped calcium aluminate glass", *J.Luminesc* 66&67, 285-289 (1996).
[11] H.Yamazaki, S.Tanabe, "Transparent Cr^{4+}-doped gehlenite($Ca_2Al_2SiO_7$) glass-ceramics for broadband amplifier", *Tech.Digest of Optical Amplifiers and their Applications 2003*, WC1, 242-244 (Otaru, July 2003).
[12] S.Tanabe, S.Todoroki, K.Hirao, N.Soga, "Phonon Sideband of Eu^{3+} in Sodium Borate Glasses", *J.Non-Cryst.Solids* **122**, 59-65 (1990).
[13] W.H.Dumbaugh, "Heavy-metal oxide glasses", *Phys.Chem.Glasses* **27**, 119 (1986).
[14] Y. Ohishi and J. Temmyo, "The Status of 1.3μm Fiber Amplifier," *Bull. Ceram. Soc. Japan* **28**, 110-14 (1993).
[15] R.D.Peacock, "The Intensities of Lanthanides *f-f* Transition"; pp.83-122 in *Structure and Bonding* , Vol. 22. Ed. by J.D. Dunitz et al., (Springer-Verlag, Berlin, 1975),
[16] W.T.Carnall, P.R.Fields, K.Rajnak, "Electronic Energy Levels in the Trivalent Lanthanide Aquo Ions. I. Pr^{3+}, Nd^{3+}, Pm^{3+}, Sm^{3+}, Dy^{3+}, Ho^{3+}, Er^{3+}, and Tm^{3+}," *J. Chem. Phys.*, **49**[10], 4424-4442 (1968).
[17] M.J.Weber, "Probabilities for Radiative and Non-radiative Decay of Er^{3+} in LaF_3", *Phys. Rev.* **157**[2] 262-72(1967).
[18] S.Tanabe, "Optical transitions of rare earth ions for amplifiers: How the local structure works in glass", *J.Non-Cryst.Solids* **259**, 1-9 (1999).
[19] M.Onishi, S.Tanabe, K.Hirao, "Spectroscopy of Er^{3+}-doped antimony silicate glasses for broab-band amplifier", *Tech.Digest of Optical Amplifiers and their Applications 2003*, WC3, 248-250 (Otaru, July 2003).
[20] K.Arai, H.Namikawa, K.Kumata, T.Honda, H.Ishii, T.Handa, "Aluminum or phosphorus co-doping effects on the fluorescence and structural properties of neodymium-doped silica glass", *J.Appl.Phys.* 59[10], 3430-36 (1986).
[21] Y.G.Choi, K.H.Kim, B.J.Park, J.Heo, "1.6μm emission from Pr^{3+}:$(^3F_3,^3F_4) \rightarrow {}^3H_4$ transition in Pr^{3+}- and Pr^{3+}/Er^{3+}-doped selenide glasses", *Appl.Phys.Lett.* 78[9], 1249-51 (2001).
[22] T.H.Lee, J.Heo, "Spectroscopic properties of Ho^{3+} doped chalcogenide glasses for 1.6μm(U-band) fiber optic amplifier", *Abstract Int'l Sympo. on Photonics Glasses 2002*, 20 (Shanghai, Oct. 2003).
[23] M.Nishi, S.Tanabe, K.Fujita, K.Hirao, *Tech.Digest of Optical Amplifiers and their Applications 2003*, WC2, 245-247 (Otaru, July 2003).
[24] E.Ishikawa, M.Nishihara, Y.Sato, C.Ohshima, Y.Sugaya, J.Kumasato, "Novel 1500-nm band EDFA with discrete Raman amplifier", *ECOC2001*, PD.1.1.2, (Amsterdam, 2001).
[25] M.Nishi, S.Tanabe, M.Inoue, M.Takahashi, K.Fujita, K.Hirao, "Fluorescence properties of Er^{3+}-doped YAG nanocrystals synthesized by glycothermal method", *J.Ceram.Soc.Jpn* (2004) in press.
[26] S. Tanabe and X. Feng, "Temperature variation of near-infrared emission from Cr^{4+} in aluminate glass for broadband telecommunication," *Appl. Phys. Lett.* **77**, 818-820 (2000).

[27] H. Eilers, U. Hommerich, S. Jacobsen, W. Yen, K. Hoffman, and W. Jia, "Spectroscopy and dynamics of Cr^{4+}:$Y_3Al_5O_{12}$," *Phys. Rev. B* 49, 15505-513 (1994).
[28] V. Petricevic, S. K. Gayen, and R. R. Alfano, "Laser action in chromium-activated forsterite for near-infrared excitation: Is Cr^{4+} the lasing ion ?", *Appl. Phys. Lett.* **53,** 2590-2592 (1988).
[29] X. Feng and S. Tanabe, "Spectroscopy and crystal-field analysis for Cr(IV) in alumino-silicate glasses", *Opt. Mater.* **20**, 63-72 (2002).
[31] P.A.Tick, N.Borrelli, L.K.Cornelius, M.A.Newhouse, "Transparent glass ceramics for 1300nm amplifier application", *J.Appl.Phys.*78[11], 6367-6374 (1996).
[32] G. H. Beall, "Transparent Cr^{4+} doped forsterite glass-ceramics for photonic applications," *Proc. Int. Congr. Glass.* Vol. 2. Extended Abstracts, 5-6 (2001).
[33] L. R. Pinckney, "Transparent β-willemite glass-ceramics", *Proc. Int. Congr. Glass.* **Vol. 2.** Extended Abstracts, 7-8 (2001).
[34] E. Munin, A. Villaverde, M. Bass, and K. C. Richardson, "Optical absorption, absorption saturation and a useful figure of merit for chromium doped glasses," *J. Phys. Chem. Solids.* **58,** 51-57 (1997).
[35] L. Merkle, T. Allik, and B. Chai, "Crystal growth and spectroscopic properties of Cr^{4+} in $Ca_2Al_2SiO_7$ and $Ca_2Ga_2SiO_7$," *Opt. Mater.* **1,** 91-100 (1992)

GRADIENT-INDEX (GRIN) LENSES AND OTHER OPTICAL ELEMENTS BY SLURRY-BASED THREE DIMENSIONAL PRINTING

Hong-Ren Wang and Michael J. Cima
Department of Materials Science and Engineering
Massachusetts Institute of Technology
77 Massachusetts Avenue
Cambridge, MA 02139

Brian D. Kernan and Emanuel M. Sachs
Department of Mechanical Engineering
Massachusetts Institute of Technology
77 Massachusetts Avenue
Cambridge, MA 02139

ABSTRACT

The slurry-based three-dimensional printing (S-3DP™) process has been used to fabricate complex ceramic structure materials by printing organic binders in selective positions of each printing layer. S-3DP™ has the advantages of short processing time and ability to control fine structures in bulk materials. This process has been modified to fabricate functionally graded materials by depositing different concentrations of dopant into selective positions. Functionally graded optical elements can be made when index-changing materials are printed. S-3DP™ provides the control of index variation in three dimensions inside bulk optical materials. Alumina-doped silica was chosen to demonstrate the fabrication of functionally graded optical elements. Aluminum nitrate, which decomposes into alumina during heat treatment, was dissolved in deionized water and printed into the silica powder bed with a maximum concentration of 2.5 mol%. The alumina-doped powder bed was dehydrated at 1000 °C for 24 hours and sintered at 1650 °C for 30 minutes in a vacuum furnace (5×10^{-6} torr). Alumina-doped silica GRIN lenses with axial and radial index variations have been successfully fabricated using S-3DP™. The distribution of alumina after sintering was measured by Electron Probe Microanalysis (EPMA) and compared with the design profile. The resulting optical effect was examined by measuring the effective focal length of the GRIN lens and compared with the theoretical value. Other optical elements, such as volume phase grating elements, will also be discussed.

INTRODUCTION

Conventional glass-based GRIN lenses, as shown in Fig. 1, have been fabricated by various methods, including molecular stuffing[1], ion exchange[2-7] and sol-gel[8,9] techniques, which rely on stuffing of base glass compositions with index altering cations. The diffusion-controlled nature of these processes results in long processing times (typically > 100 hours), thereby limiting feasible component sizes to less than 13 mm. A comparison of the lens diameter and the index gradient difference (Δn) of radial GRIN rods prepared by various methods is provided in Fig. 1[8]. The maximum index gradient difference currently produced, i.e., $\Delta n < 0.2$, is limited not only by the base glass composition, but the dopant concentration profiles achievable by these methods. Commercially available SELFOC® lenses, for example, prepared by ion exchange exhibit a maximum Δn value of 0.124 for components ranging in size from 1.0 to 4.5 mm[10]. GRIN materials fabricated by ion exchange techniques are also not suitable for high temperature applications because the migration of alkali ions results in the distortion of the index profile[8]. Alternative materials systems or fabrication methods for large-scale GRIN components with desired optical characteristics and good environmental and thermal stability are needed.

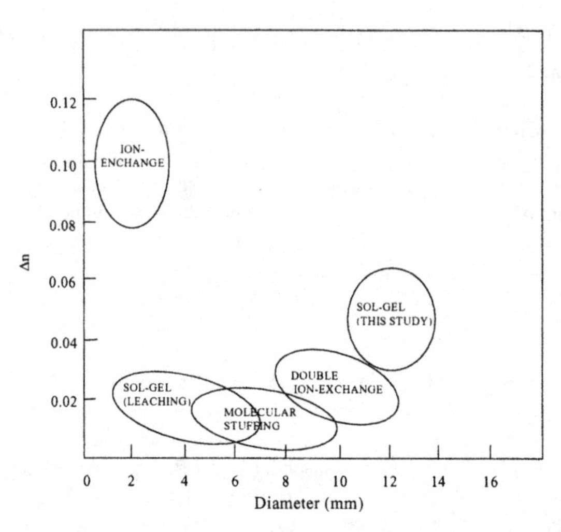

Fig. 1. Comparison of the lens diameter and Δn of radial GRIN rods prepared by various methods. [8]

Microlens arrays have been used for the interconnection between optical fibers. The optical signal is refocused when passing through individual microlens. Commercially available microlens arrays are typically made of polymer-based or glass-based materials, such as UV curable polymer and borosilicate. The focusing powder of these microlens arrays results from the curvature of individual microlens except the SELFOC® microlens array that consists of an array of GRIN lenses. The focusing powder and the lens-to-lens distance determine the performance of a microlens array. Volume phase gratings are used as filters in various applications such as Raman spectroscopy[11] and confocal microscope[12]. It is known that volume phase gratings provide higher light transmission than the traditional amplitude gratings. Volume phase gratings are typically made by shining diffraction pattern generated from two laser sources onto photosensitive materials such as dichromated gelatin and iron-doped lithium niobate. The size and depth of the grating are generally limited by the powder of laser and the photosensitive materials.

The S-3DP™ technology[13-15] is an agile, facile method of producing near-net shape advanced ceramic components. Parts are constructed in a layer-by-layer build sequence. Each powder bed layer is created by jetting a ceramic slurry onto a substrate. The as-cast layer is then dried and a binder, which cements the ceramic particles, is selectively deposited in the desired pattern. The excess powder is then removed to produce a three dimensional structure. The S-3DP™ process can be modified to fabricate functionally graded materials, such as GRIN lenses, by depositing different amounts of dopant instead of a binder into each layer. The S-3DP™ processes for vertical and radial GRIN lenses are shown schematically in Fig. 2(a) and Fig. 2(b), respectively. Production of GRIN materials by S-3DP™ offers several advantages over conventional processing methods, including reduced processing times (< 70 hours) yielding economical fabrication of large-scale components, improved compositional flexibility, and increased index profile dimensionality. The improved flexibility and compositional control offered by S-3DP™

results in a single component lens system with greater functionality. The lens stacking required to overcome optical aberrations of a photographic lens system can be eliminated by taking advantages of the additional degree of freedom offered by S-3DP™ GRIN lenses. This research attempts to demonstrate the concept of utilizing S-3DP™ to the fabrication of larger scale GRIN lenses and other optical elements, including micro lens arrays and optical volume phase gratings. Several silica-based material systems have been considered and tested. The alumina-doped silica glass was chosen as the material system.

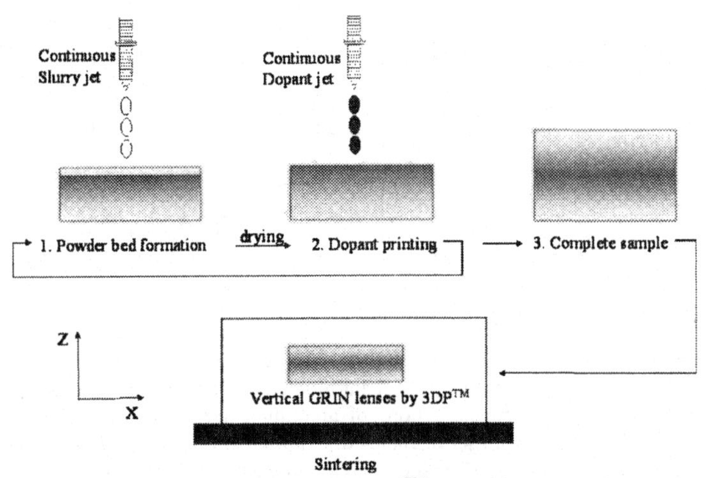

Fig. 2(a). The schematic drawing of the S-3DP™ process for vertical GRIN lenses.

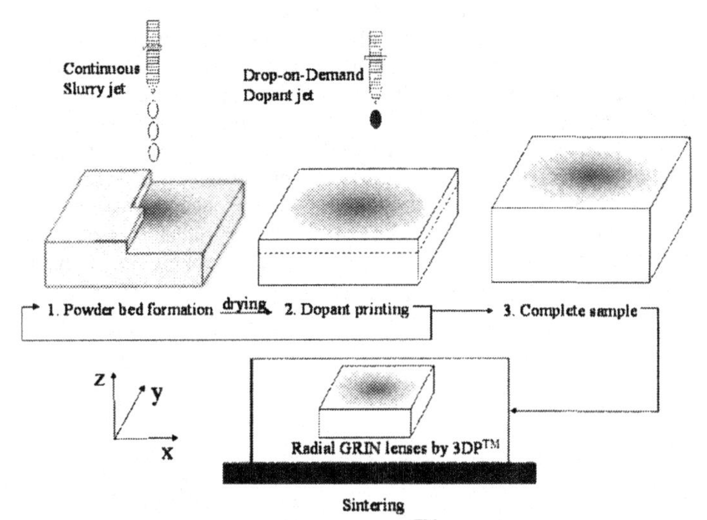

Fig. 2(b). The schematic drawing of the S-3DP™ process for radial GRIN lenses.

EXPERIMENTS

The amorphous silica powder (Mitsubishi Chemical Company) used in this research had a median particle size of 1.4 μm and a surface area of 2.666 m^2/g. Aluminum nitrate nanohydrate ($Al(NO_3)_3*9H_2O$, Alfa Aesar) was used as the dopant source. Vertical index variation GRIN lenses were fabricated from a 30 vol% silica slurry, while a 22.5 vol% slurry was used to fabricate the radial index variation GRIN lens. Both slurries were ball-milled with glass media for 20 hours before printing. The chemical compositions of the slurries are shown in Table I. Boric Acid was added to lower the sintering temperature.

Table I. The chemical compositions of the silica slurries.

Silica Powder (vol%)	Deionized Water (vol%)	Methanol (vol%)	2-Propanol (vol%)	Poly (ethylene glycol) (MW:400)	TMAH* (M)	NH4OH (M)	H3BO3 (wt%)
30	35	35	0	3 wt% based on silica	0	0.20	1 wt% based on silica
22.5	38.75	0	38.75	0	0.063	0	1 wt% based on silica

*tetramethylammonium hydroxide

Aluminum nitrate-doped silica powder beds with vertical compositional variation were fabricated by S-3DP™ using a continuous dopant jet. The vertical compositional variation was achieved by printing different concentrations of aluminum nitrate solution while keeping the flow rate of the continuous dopant jet unchanged. The thickness of each printed layer was 73 μm. Two vertical concentration profiles with maximum concentration in the center were printed to create two vertical GRIN lenses with different optical effects, shown in Fig. 3. Each printed slurry layer and dopant layer was dried at 65 ^{0}C for 50 seconds and 55 layers were printed. Aluminum nitrate-doped silica powder beds with radial compositional variation were made with a 3DP™ machine equipped with Drop-on-Demand (DoD) printing nozzles, which allow the dopant solution to be deposited in selective region drop by drop. The drop size of the aluminum nitrate solution was 45 μm. The thickness of each dried slurry layer was 40 μm. The designed concentration profile is shown in Fig. 4. Each printed slurry layer was dried at 100 ^{0}C for 50 seconds while each printed dopant layer was dried in a microwave oven for 1 minute. Microlens arrays were made by printing individual dopant drops into a silica powder bed with a drop-to-drop distance of 120 μm. One layer of the array pattern was printed. Two line patterns of the volume phase grating samples were also printed using DoD printing nozzles with the printing line spacing of 120 μm and 160 μm, respectively.

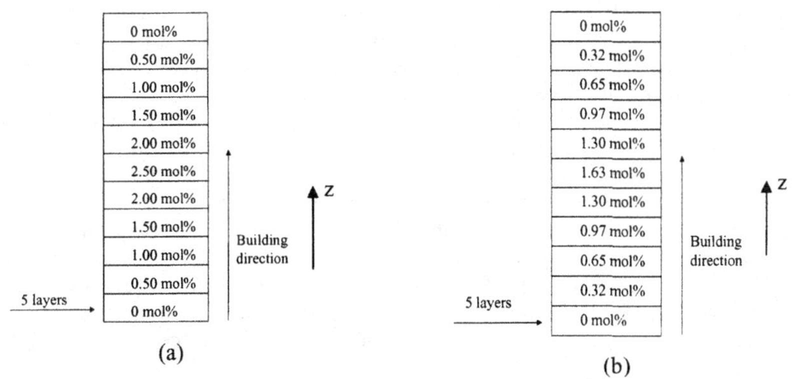

Fig. 3. The concentration profiles of alumina in the GRIN lenses of (a) Design 2.5% max and (b) Design 1.63% max.

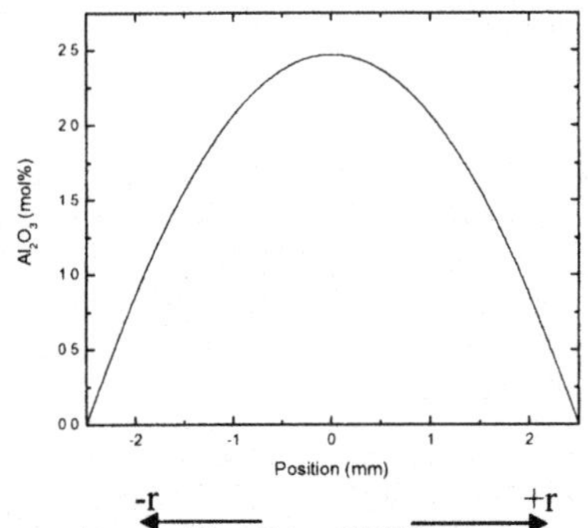

Fig. 4. The concentration profile of the GRIN lens with radial index variation

The aluminum nitrate-doped silica powder beds were heated at a temperature of 900°C for 4 hours in air to remove the hydroxyl groups introduced by the aluminum nitrate solution as well as the organic additives. Sintering was performed in a vacuum furnace (Centoor, MRF, pressure ~ 5×10^{-6} torr) at various temperatures, holding times, and cooling rates. It was found that un-doped silica powder beds were sintered into optical transparency at 1500°C for 30 minutes, as shown in Fig. 5. An additional dehydration process at 1000 °C for 24 hours in the vacuum furnace for powder beds with a maximum alumina concentration of 2.50 mol% or higher was required to completely remove the residual hydroxyl groups that form bubbles during sintering.

Fig. 5. The transparent un-doped silica powder bed sintered at 1500 °C for 30 minutes.

X-ray Diffraction (XRD) was used to detect crystallization of the doped powder bed after heat treatment. The sintered powder beds were polished and observed under an optical microscope. Chemical compositions of the doped powder beds were measured by electron probe microanalysis (EPMA, JOEL Superprobe 733).

RESULTS

The mixture of alumina and silica tends to form mullite ($3Al_2O_3$+$2SiO_2$) at temperature higher than 950 °C[14]. The formation of mullite can be minimized by increasing the cooling rate and using the alumina concentration lower than 5 mol%[16]. The maximum alumina concentration in this study was 2.50 mol% (Fig. 4). XRD result shows no crystallization in the alumina-silica powder beds that were treated at 900 °C for 4 hours. Several sintering conditions were tested. Optical transparency was achieved for the alumina-doped powder beds by sintering at 1650 °C for 30 minutes and cooling at the maximum rate (~500 °C/minute from the sintering temperature) allowed by the furnace.

The magnifying effects of the sintered powder beds with vertical compositional variation are shown in Fig. 6. The MIT markers under the sintered powder beds are magnified in the vertical direction, as expected from the dopant concentration profile shown in Fig. 3. The object and image sizes in the vertical direction were measured, allowing the effective focal length (f_{eff}) to be determined by the following equation[17]:

$$\frac{1}{f_{\mathrm{eff}}} = \frac{1}{S_1} - \frac{H_1}{H_2 * S_1} \tag{1}$$

where H_1 is the object size, H_2 is the image size, and S_1 is the distance between the lens and the object. The effective focal lengths of the powder beds with vertical compositional variation were calculated to be 10 cm and 6.1 cm for Design 1.63% max and Design 2.50 max, respectively. The magnifying effect of the sintered powder bed with radial compositional variation is shown in Fig. 7. The effective focal lengths were also calculated to be 63.75 cm in x-direction and 52.50 cm in y-direction using the same method.

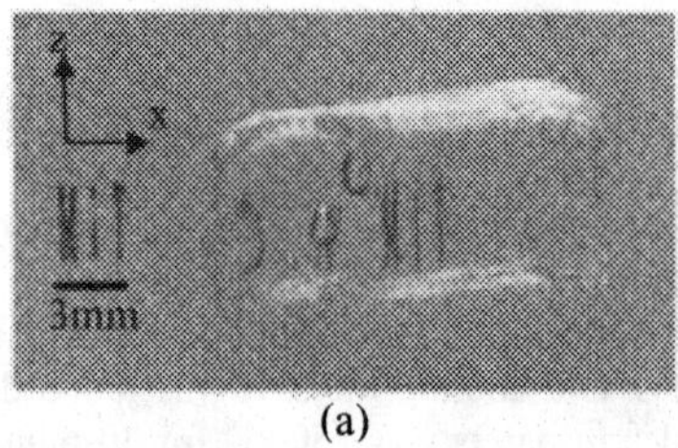

(a)

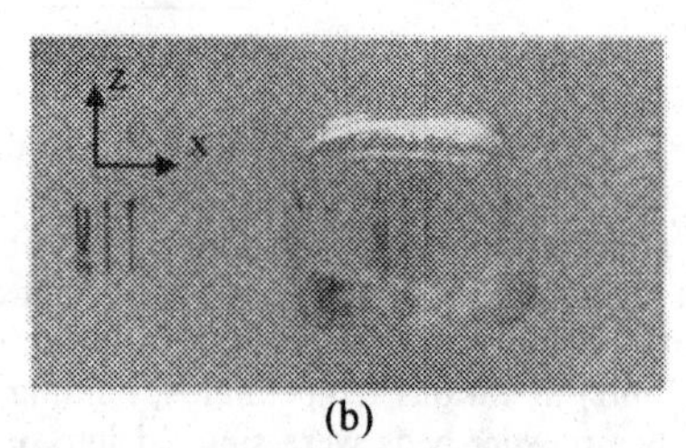

(b)

Fig. 6. The vertical enlargement with the alumina-doped GRIN lenses above an MIT marker. (a)Maximum alumina concentration: 1.63 mol%, (b)Maximum allummina concentration: 2.50 mol%

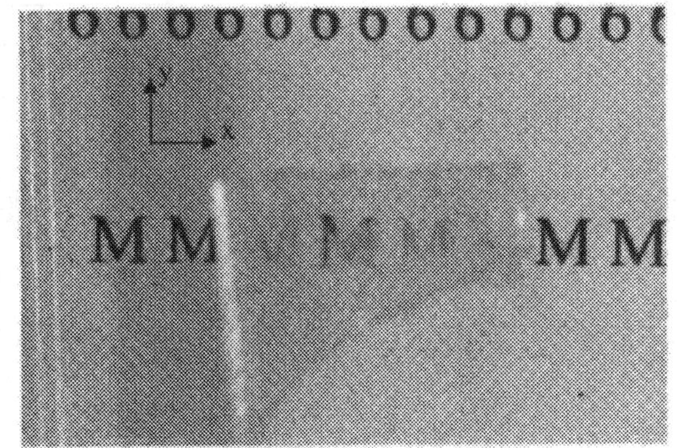

Fig. 7. The enlargement with the radial alumina-doped GRIN lenses.

The chemical composition profiles of the sintered powder beds with vertical compositional variation measured by EPMA are shown in Fig. 8 while the chemical composition profiles in x-direction and y-direction of the sintered powder bed with radial compositional variation are shown in Fig. 9. The EPMA mapping result of the sintered microlens array is shown in Fig. 10. The average microlens size can be measured from Fig. 10 to be about 77 μm while the dot-to-dot distance in both x-direction and y-direction is about 90 μm. Fig. 11 shows the EPMA results of the sintered volume phase gratings. The average line spacing after sintering can be measured from Fig. 11 to be 135 μm and 104 μm, respectively.

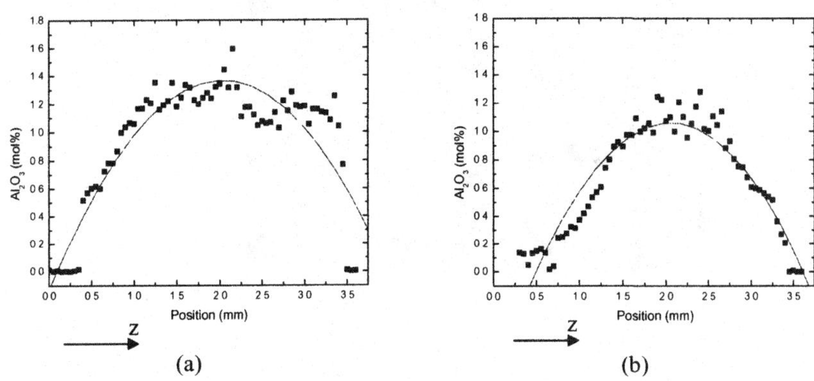

Fig. 8. The dopant distribution of the vertical compositional variation GRIN lenses of (a) Design 2.5% max and (b) Design 1.63% max

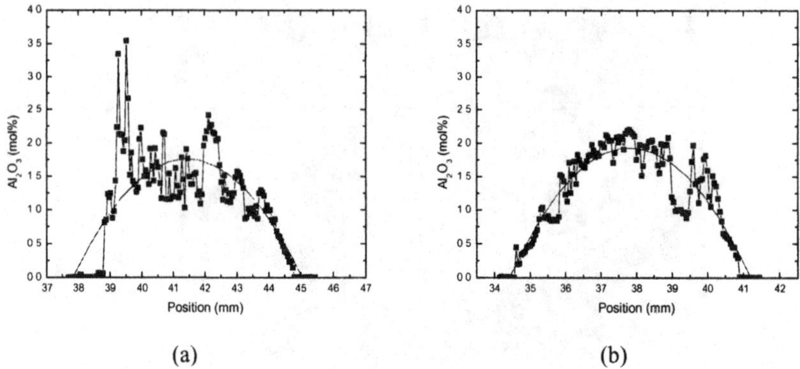

(a) (b)

Fig. 9. The dopant distributions of the GRIN with radial index variation in (a) x-direction and (b) y-direction.

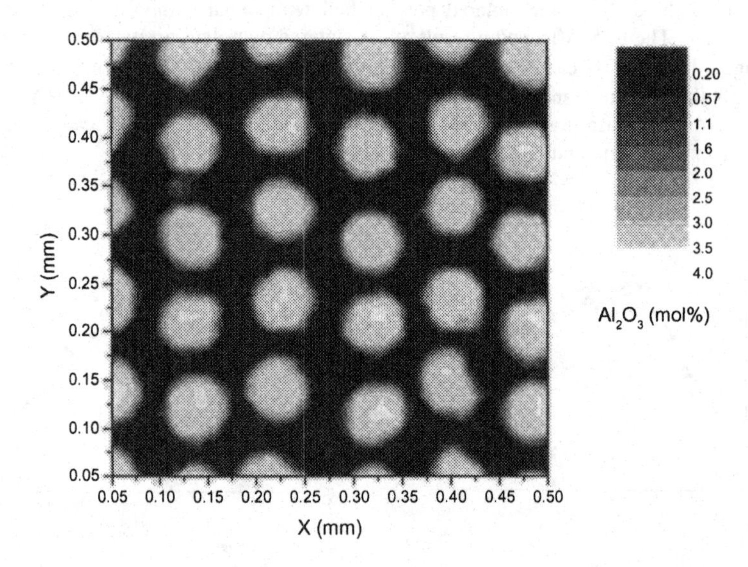

Fig. 10. The 2-dimensional EPMA result of a sintered microlens array.

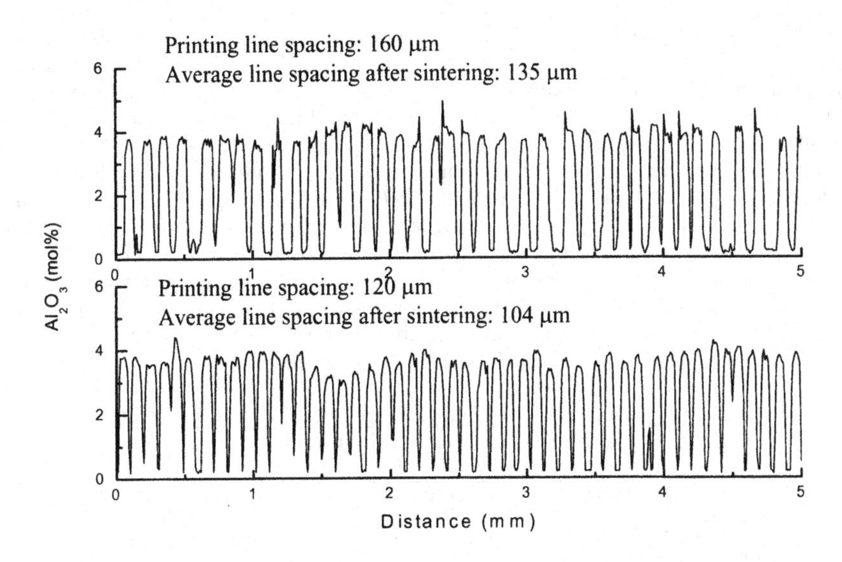

Fig. 11. The EPMA results of the sintered volume phase gratings.

DISCUSSION

The theoretical focal length (f_{th}) of a GRIN slab with a parabolic index of refraction profile is given by the following equation[18]:

$$f_{th} = \frac{1}{\left(\frac{n_{max}^2 - n_{min}^2}{0.25\,w^2}\right)^{1/2} \sin\left(\frac{d}{0.5w}\left(1 - \frac{n_{min}^2}{n_{max}^2}\right)^{1/2}\right)} \tag{2}$$

where w is the width of the GRIN slab, d is the thickness of the GRIN slab, n_{min} is the minimum index of refraction, and n_{max} is the maximum index of refraction. No direct measurement of index of refraction has been made in this study. The index of refraction, n, of fused silicate containing alumina, however, has been found to vary linearly with the alumina content, M, as described by Equation 3[19]:

$$n=1.4580+0.00192M \tag{3}$$

where M is the alumina concentration in mol%. The alumina concentration profiles of the Design 1.63% max and Design 2.5% max samples are shown in Fig. 8. The profiles are fitted with parabolic curves. The maximum alumina concentrations of the Design 1.63% max and Design 2.5% max samples are found to be 1.04 mol% and 1.35 mol% and the maximum indices of refraction (n_{max}) of the samples are calculated to be 1.46 and 1.4606 from Equation (3). The theoretical focal lengths of the sintered powder beds, assuming a parabolic index profile, are then calculated and compared with the effective focal lengths, as shown in Table II. The effective focal length is close to the theoretical value for the Design 1.63% max sample. The relatively larger difference between the effective and theoretical value of the Design 2.5% max sample is due to the fact that the concentration profile is not ideally parabolic. The alumina concentration profiles of the sintered powder bed with radial compositional variation in x-direction and y-direction, shown in Fig. 9, are also fitted with parabolic curves. The maximum alumina concentrations in x-direction and y-direction on the parabolic curves are found to have values about 1.70 mol% and 1.92 mol%, respectively. The theoretical focal lengths of the radial GRIN lens in x-direction and y-direction can also be calculated by replacing the width (w) of the GRIN slab with the diameter of the radial GRIN lens in Equation 2. The theoretical focal lengths are then calculated to be 52.11 cm in x-direction and 50.77 cm in y-direction from Equation 2, as also shown in Table II. The effective focal length in x-direction (63.75 cm) has about 13 cm difference comparing with the value of the theoretical focal length in x-direction (50.77 cm) calculated previously. The difference can be attributed to the deviation of the concentration profile in x-direction from a parabolic curve, as shown in Fig. 9. The effective focal length in y-direction, however, is close to the value of the effective focal length (52.50 cm) due to the fact that the concentration profile in y-direction is fitted well to the parabolic curve.

Table II. The effective (f_{eff}) and theoretical (f_{th}) focal lengths of different GRIN lenses

	Thickness, d (cm)	*Width(diameter)*, w (cm)	f_{eff}(cm)	f_{th} (cm)
Design 1.63% max	0.55	0.30	10.00	10.30
Design 2.5% max	0.60	0.30	6.10	7.28
Radial GRIN lens, x-direction	0.27	0.60	63.75	52.11
Radial GRIN lens, y-direction	0.27	0.63	52.50	50.77

It is noted that the length scale of the concentration deviation in both vertical and radial compositional GRIN lenses is in the scale of millimeters, as shown in Fig. 8 and Fig. 9. Davis and Pask studied the diffusion in the alumina-silica system. Their result showed that the diffusivity of alumina in silica at 1650 °C is between 2×10^{-11} cm^2/s (2.44 mol% Al_2O_3) to 1×10^{-10} cm^2/s (5.35 mol% Al_2O_3)[20]. It is thus unlikely that the deviation is caused by the diffusion of aluminum during sintering at 1650 °C for 30 minutes. Migration of aluminum nitrate solution during the printing process is believed to cause the concentration deviation. Differential slip casting, which occurred during the printing process, is resulted from the different slip casting rates of the powder bed at different positions. The presence of dried dopant within the pore space lowers the slip casting rate and slurry can migrate from slow casting regions to fast casting regions. The slow casting region can have a higher green density than the fast casting region. The capillary force difference between pores then causes the as-deposited aluminum nitrate solution to migrate within the powder bed. This results in the concentration deviation in the concentration variation plane. The asymmetric concentration profile in x-direction of the radial GRIN lens, as shown in Fig. 9(a), is believed to be resulted from the differential slip casting caused by the asymmetric deposition of slurry. The direction of slurry deposition is along x-direction and is symmetric in y-direction. This problem could be solved by alternating the direction of slurry deposition every layer. More experiments will be done to understand and solve the problem in the future.

The theoretical focal length of each microlens can be estimated using Equation 2 and the chemical analysis result from Fig. 10. The maximum alumina concentration is about 3.5 mol% from Fig. 10 and results an index of refraction of 1.4647 from Equation 3. The lens diameter is

77 μm as mentioned previously. The lens thickness is 40 μm since only one layer of microlens array pattern was printed. The theoretical focal length can then be calculated to be 0.28 cm. Shorter focal length can be achieved by increasing the number of printed layers of microlens array pattern. The theoretical focal length is calculated to be 276 μm for a micronlens with the same alumina concentration and lens diameter but with lens thickness of 600 μm. The focal length of each microlens in a micrlens array can be changed as desired since the lens thickness of each microlens can be controlled individually by S-3DP™.

A He-Ne laser source with wavelength of 633 nm was shined perpendicularly to the gratings. Diffraction patterns were observed at a distance of 2.8 m from the gratings. The average dot-to-dot distance in the diffraction pattern of the grating with 135 μm line spacing is 1.44 cm while the average dot-to-dot distance in the diffraction pattern of the grating with 104 μm line spacing is 1.80 cm, as shown in Fig. 12. The results are consistent with the fact that the dot-to-dot distance in the diffraction pattern of a phase grating is inversely proportional to its line spacing when the laser wavelength and observing distance are the same[21].

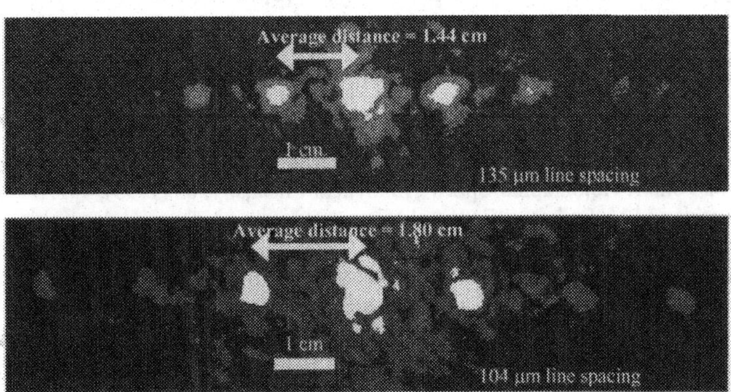

Fig. 12. Diffraction patterns of the volume phase gratings with line spacing of 135 μm and 104 μm, respectively.

CONCLUSION

The alumina-silica system has been studied for the fabrication of GRIN lenses by S-3DP™. Optically transparent alumina-silica GRIN lenses were obtained by sintering at 1650 °C for 30 minutes. GRIN lenses with vertical and radial compositional variations have been successfully fabricated. EPMA results show the deviation of the actual dopant concentration from the designed value. It is believed to be resulted from the different slip casting rates between the doped and un-doped regions in the powder beds. A microlens lens array with a theoretical focal length of 0.28 cm is fabricated as well as the volume phase gratings with line spacing of 104 μm and 135 μm, respectively.

ACKNOWLEDGEMENT

The authors would like to thank Christopher C. Stratton for his help with 3DP™ file generation.

REFERENCES
[1]J. H. Simmons and *et al.*, "Optical Porperties of Waveguides Made by Porous Glass Process," *Applied Optics*, **18** [16] 2732-2733 (1979).
[2]I. Kitano and *et al.*, "A Light Focusing Fiber Guide Prepared by Ion-exchange Techniques," *J. Japan Soc. App. Phys.*, **39**, 63-70 (1970).
[3]J. E. Samuels and D.T. Moore, "Gradient-index Profile Control from Molten Salt Baths," *Applied Optics*, **29** [28] 4042-4050 (1990).
[4]S. Ohmi and *et al.*, "Gradient-index Rod Lens Made by a Double Ion-exchange Process," *Applied Optics*, **27** [3] 496-499 (1988).
[5]S. N. Houde-Walter and D.T. Moore, "Delta-n Control in GRIN glass by Additives in AgCl Diffusion Baths," *Applied Optics*, **25** [19] 3373-3378 (1986).
[6]A. R. Cooper and M.A. el-Leil, "Index Variation from Field-assisted Ion Exchange," *Applied Optics*, **19** [7] 1087-1091 (1980).
[7]A. D. Pearson, W.G. French, and E.G. Rawson, "Perparation of a Light Focusing Glass Rod by Ion-exchange Techniques," *Applied Physics Letters*, **15** [2] 76-77 (1969).
[8]M. Yamane and *et al.*, "Graded Index Materials by the Sol-gel Process," SPIE Optical Engineering Press , 1993.
[9]T. M. Che, J.B. Caldwell, and R.M. Mininni, "Sol-gel Derived Gradient Index Opitcal Materials," *Sol-Gel Optics*, **1328,** 145-159 (1990).
[10]D. T. Moore, "Gradient-Index Optics: A Review," *Applied Optics*, **19** [7] 1035-1038 (1980).
[11] J. M. Tedesco, H. Owen, D. M. Pallister, and M. D. Morris "Principles and Spectroscopic Applications of Volume Holographic Optics", *Analytical Chemistry*, **65** (1993), 441A.
[12]G. Barbastathis and M. Balberg, "Confocal Microscopy with a Volume Holographic Filter", *Optics Letters*, **24** (1999), 811.
[13]E. M. Sachs, M. J. Cima, P. Williams, D. Brancazio and J. Cornie, *J. Eng. Ind.*, **114** (1992) 481.
[14]J. E. Grau, J. Moon, S. Uhland, M. J. Cima and E. M. Sachs, pp. 317-79 in Proceedings of the Solid Freeform Fabrication Symposium, edited by J. J. Beaman, H. L. Marcus, D. L. Bourell, J. W. Barlow and T. Crawford, University of Texas, Austin, TX, 1997.
[15]M. J. Cima, M. Oliveria, H.-R. Wang, E. M. Sachs, and R. Holman, pp. 216-23 in Proceedings of the Solid Freeform Fabrication Symposium, edited by J. J. Beaman, H. L. Marcus, D. L. Bourell, J. W. Barlow and T. Crawford, University of Texas, Austin, TX, 2001.
[16]J. F. MacDowell and G. H. Beall, "Immiscibility and Crystallization in Al_2O_3-SiO_2 Glasses," *J. Am. Ceram. Soc.*, **52** [1] 17-25 (1969).
[17]E. Hecht, "Optics", 4th edition, Addison-Wesley, 2002, p. 159.
[18]B. E. A. Saleh and M. C. Teich, "Fundamentals of Photonics", John Wiley & Sons Inc., New York, 1991, p. 23.
[19]K. Nassau, J. W. Shiever, and J. T. Krause, "Preparation and Properties of Fused Silica Containing Alumina," *J. Am. Ceram. Soc.*, **58** [9-10] 461 (1975).
[20]R. F. Davis and J. A. Pask, *J. Am. Ceram. Soc.*, **55** (1972) 525.
[21]J. W. Goodman, "Introduction to Fourier Optics", 2nd edition, McGraw-Hill, 1996, p.81-83.

NON-SILICA MICROSTRUCTURED OPTICAL FIBERS

T.M. Monro, H. Ebendorff-Heidepriem, X. Feng
Optoelectronics Research Centre
University of Southampton
Southampton SO171BJ, United Kingdom

ABSTRACT

Microstructured fibers offer a range of optical properties that cannot be achieved in conventional (solid) optical fibers. Although to date, most work in this field has focused on silica glass, non-silica microstructured fibers are attractive for applications including nonlinear devices, mid-IR transmission and photonic bandgap operation. A review of progress in the development and applications of non-silica microstructured fibers will be presented.

INTRODUCTION

Research in the field of microstructured optical fibers has progressed rapidly in recent years because these fibers can offer a diverse range of novel optical properties that cannot be provided by more conventional optical fibers, which opens up diverse potential applications in telecommunications, industrial processing, metrology, medicine and beyond. The transverse cross section of a microstructured fiber typically contains an approximately periodic arrangement of air holes that run along the fiber length. The vast majority of microstructured fibers that have been produced to date are air/silica fibers that contain an arrangement of air holes embedded in undoped silica glass [1-4].

The holes within the microstructured fibers act as the fiber cladding, and light can be guided using two quite different mechanisms. *Index-guiding* microstructured fibers guide light due to the principle of modified total internal reflection, and such fibers are widely known as holey fibers (HFs) [2]. In this class of fibers, the holes effectively act to lower the effective refractive index in the cladding region, and so light is confined within the solid core, which has a relatively higher refractive index. This guidance mechanism does not rely on the use of a periodic arrangement of holes [5]. However, since in practice silica holey fibers are generally made by stacking a large number of circular capillaries, the resulting air holes are typically arranged on a hexagonal lattice, and so these fibers are also often referred to as photonic crystal fibers [3].

Index-guiding holey fibers exhibit novel optical properties for two principal reasons: (1) the effective refractive index of the holey cladding region can vary strongly as a function of the wavelength of light guided by the fiber and (2) the air/glass features provide a large refractive index contrast. In addition, the optical properties of index-guiding HFs are determined by the configuration of air holes used to form the cladding, and many different arrangements can be envisaged within this flexible fiber geometry. For these reasons, it is possible to design single material silica fibers with properties that are not possible in conventional fibers such as broadband single-mode guidance [6] or anomalous dispersion in single mode fibers at wavelengths down to 560 nm [7]. Simply by scaling the dimensions of the features within the fiber profile, pure silica HFs can have mode areas ranging over three orders of magnitude. At one extreme, HFs with large mode area are of great promise for high power transmission systems (see for example [8], [9]). At the opposite end of the scale, by combining small-scale cladding features with the large effective index contrast possible in HFs, it is possible to achieve tight mode confinement, and thus high effective fiber nonlinearities can be obtained. Such fibers have been used in a variety of nonlinear devices including devices for wavelength conversion, supercontinuum generation, all-optical switching and data regeneration, pulse compression and soliton formation (see for example [10]-[21]).

Photonic bandgap microstructured fibers (PBGFs) can guide light by an alternative guidance mechanism, the so called photonic bandgap effect, if the air holes that define the cladding are arranged on a periodic lattice [4]. By suitably breaking the periodicity of the cladding it is possible to introduce a localized mode within this defect. Such a defect can act as a core, and guide light within well-defined frequency windows. Hollow core photonic bandgap fibers can be designed to transmit at near-infrared wavelengths and are of particular interest for gas/light interactions and the transport of high intensity optical fields.

In the single-material holey and photonic bandgap fibers described thus far, light is solely confined by the holes in the cladding. Hybrid microstructured fibers are another class of fiber that combines a high-index doped core with a holey cladding. At one extreme, in *hole-assisted fibers*, light is guided by the relatively higher index of the doped core, and the air holes located in the cladding of a conventional solid fiber act to subtly modify properties such as dispersion [22]. In air-clad fibers, an outer cladding with a high air-filling fraction creates a high numerical aperture inner cladding which allows the realization of cladding-pumped high power lasers [23]. Alternatively, dopants can be added to the core of a HF to create novel HF-based amplifiers and lasers (see for example [24]).

Most work to date in the field of microstructured fibers has focused on silica glass technology, which allows the definition of a diverse range of high quality transverse fiber cross-sections via the capillary stacking fabrication technique. This field has now begun to come of age. Advances in fiber fabrication techniques have reduced the losses of both fiber types dramatically in recent years (to 0.3dB/km for HFs [25] and 1.7 dB/km for PBGFs [26]). The prospect of microstructured silica fibers with a lower transmission loss than conventional fibers represents a tantalizing possibility with the potential to revolutionize telecommunications.

The combination of the microstructured fiber concept with non-silica glasses is a relatively unexplored field which promises the development of a host of new fibers and operational regimes not achievable in existing fibers. For example, recently compound glasses with high intrinsic optical nonlinearity (such as lead-silicate, tellurite and bismuth-oxide) have been used to produce HFs with extreme values of effective fiber nonlinearity [27]-[34]. In addition, polymer-based HFs have been reported [35]. A number of non-silica glasses transmit at wavelengths substantially beyond silica into the mid-IR (for example chalcogenide glasses), and microstructured fibers made from such materials promise a new means of providing power delivery at these wavelengths. Note that in contrast to conventional fibers, microstructured fibers can be made from a single material, which eliminates the problems induced by the requirement that the core/cladding glasses are thermally and chemically matched. This opens up the prospect of using microstructured fiber technology as a tool for realizing optical fibers from an extremely broad range of optical materials.

Other classes of microstructured fiber have also begun to emerge. In particular, microstructured fibers with solid cladding designs have recently been demonstrated. In these fibers, low index glass inclusions are embedded in a higher-index glass matrix either in the form of circular inclusions [36] or in nested layers [37]. These fibers combine the practicality of a solid cladding with the design flexibility provided by the transverse microstructure.

The aim of this paper is to provide an overview of the progress to date and range of potential applications for non-silica microstructured optical fibers. We begin by reviewing the fabrication techniques that have been employed to make non-silica microstructured fibers.

FABRICATION APPROACHES

Preform Fabrication

As noted above, the vast majority of microstructured fibers that have been produced to date have been made from silica glass, and the preforms for these fibers are usually fabricated using stacking techniques. Capillary tubes are stacked in a hexagonal configuration, and the central capillary can then be replaced with a solid glass rod, which ultimately forms the fiber core. The stacking procedure is flexible: for example active fibers can be made using rare-earth-doped core rods, off-center or multiple core fibers can be readily made by replacing a non-central or multiple capillaries, etc. The reproducibility of the structured profile in the preform depends on having long uniform capillaries to stack and achieving a good stack with many tubes. One significant drawback of the stacking approach is that the preform fabrication is labor-intensive.

Recently, attention has focused on the fabrication of microstructured fibers in a range of non-silica glasses. In general, such compound glasses have low softening temperatures relative to silica glass, which allows new techniques to be used for the fabrication of structured preforms such as rotational casting and extrusion. The technique that has been used for the majority of non-silica microstructured fibers made thus far is extrusion [27-34]. In this process, a glass billet is forced through a die at elevated temperatures near the softening point, whereby the die orifice determines the preform geometry. Once the optimum die geometry and process parameters have been established, the preform fabrication process can be automated. In this way good reproducibility in the preform geometry has been achieved [32]. One advantage of the extrusion technique is that the preform for the microstructured part of a fiber can be produced in one step. In addition, extrusion allows access to a more diverse range of cladding structures, since the holes are not restricted to hexagonal arrangements.

Fiber Fabrication

To produce fibers with relatively large scale features, fibers are in general drawn directly in a single step from the preform. For small scale features, the preform is first reduced to a cane of 1-2mm diameter on a drawing tower, and in a second step this cane is inserted in a solid jacketing tube and then drawn to the final fiber. One of the most challenging aspects of air/glass microstructured fiber fabrication is to prevent collapse of the holes and to achieve the target hole size and shape during caning and fiber drawing. The microstructured profile can be affected by the pressure inside the holes, surface tension of the glass and temperature gradient in the preform. Note that compared with silica glass, most non-silica glasses have significantly steeper viscosity curves, which leads to greater demands on the process control during fiber drawing. Nevertheless, a high degree in the reproducibility of the HF geometry has already been demonstrated for both lead silicate glass [32] and bismuth glass [33,34].

HIGH NONLINEARITY FIBERS

Design Concept

Microstructured fiber technology has enabled significant progress in the development of fibers with high effective nonlinearity. This is achieved by combining a small core design with a high numerical aperture (NA) to yield tight mode confinement. One common measure of fiber nonlinearity is the effective nonlinearity γ [38]: $\gamma = 2\pi n_2 / (\lambda A_{eff})$, where n_2 is the nonlinear coefficient of the material, λ the wavelength and A_{eff} is the effective mode area.

Pure silica holey fibers with effective mode areas as small as 1.3 μm^2 have successfully been fabricated, exhibiting effective nonlinearity coefficients as high as 70 $W^{-1}km^{-1}$ at 1550 nm [39], i.e. around 70 times more nonlinear than standard single mode fibers. This value represents the theoretical limit in nonlinearity that can be achieved in a silica/air fiber. However, the material nonlinearity of compound glasses [40] can be more than one order of magnitude larger than that of silica [38]. The combination of highly nonlinear glass composition and small core/high NA HF geometry allows a further dramatic increase of the fiber nonlinearity.

As an example, we estimate the magnitude of the nonlinearity that could ultimately be achieved in a bismuth oxide glass holey fiber by considering the case of a rod of glass suspended in air (see description of bismuth glass to follow). This simplified geometry represents the fundamental limit in the mode area/nonlinearity that can be achieved in an air/glass microstructured fiber made from this material. The highest nonlinearity of $\gamma = 2200$ W^{-1} km^{-1} is obtained when the core rod is ~0.8 μm in diameter (Figure 1). Note that this nonlinearity is a factor of two higher than the record nonlinearity achieved thus far for a chalcogenide glass fiber with conventional solid cladding [41].

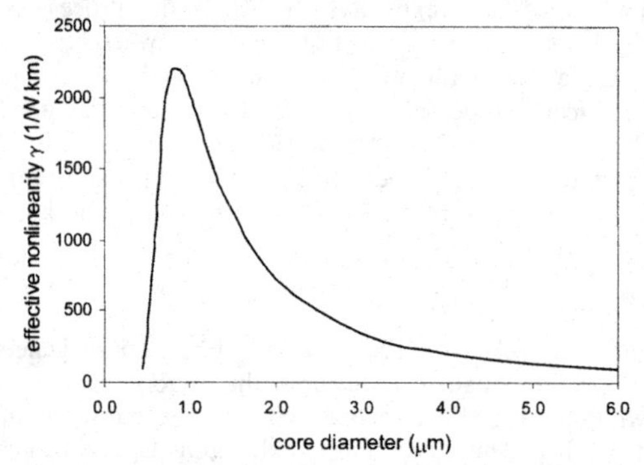

Figure 1. Calculated nonlinearity (γ) for an air-suspended rod of bismuth oxide glass

Here we review fabrication techniques that have been used to make these fibers, the range of glasses that have thus far been successfully used, and some key results that have been obtained using nonlinear non-silica holey fibers.

Fabrication Approach

All of the small-core non-silica microstructured fibers produced to date have exploited extrusion techniques to fabricate structured fiber preforms with mm-scale features from bulk glass billets. As an example, consider the structured preform shown in Figure 2(a). In this geometry, the core (center) is attached to three long fine supporting struts. The outer diameter of this preform is approximately 16mm. The preform was reduced in scale on a fiber drawing tower to a cane of approximately 1.7mm diameter (Figure 2(b)).In the last step, the cane was inserted within an extruded jacket tube, and this assembly was drawn to the final fiber (Fig.1(c)). The illustrative examples shown in figures 2 (a)-(c) are made from the lead silicate glass SF57, and are taken from Reference [32]. This procedure has also been used to make nonlinear fibers from other materials including bismuth glass (for more detail see below). A similar technique has been used to produce small-core tellurite microstructured fibers in which the core is supported by six long fine struts [29]. In Ref [29], the preform extrusion is performed directly on a fiber drawing tower, which allows the preform to be reduced to $\approx$1mm diameter cane directly at the time of extrusion.

The core diameter can be adjusted during fiber drawing by an appropriate choice of the external fiber diameter. Small core dimensions are chosen to provide tight mode confinement and thus high effective nonlinearity. For fiber designs of the type shown shown in figure 2, core diameters in the range 1.7-2.3 μm correspond to struts that are typically > 5 μm long and < 250 nm thick. These long, thin struts act to isolate the core optically from the external environment, and thus ensure that confinement loss is negligible [42]. Note that these fibers have an improved structure when compared with the earliest fibers of this type that were made [27]. Excellent structural reproducibility has been demonstrated using this fabrication technique [32]. Next, we review the glasses in which nonlinear microstructured fibers have successfully been fabricated to date.

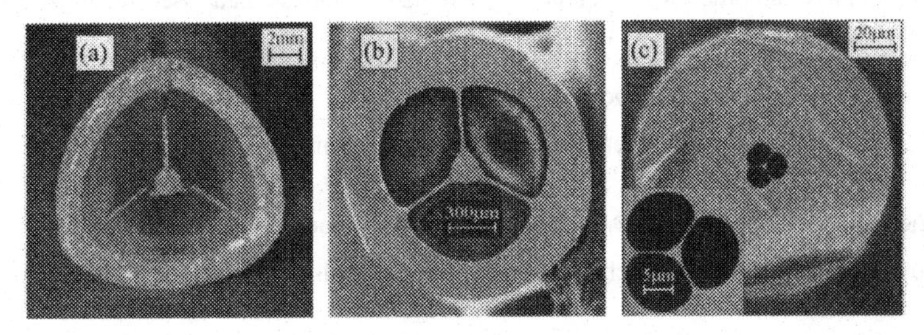

Figure 2: (a) Cross section through extruded preform, (b) SEM image of cane cross section, (c) SEM image of holey fiber cross section.

SF57: Lead silicate glasses are a promising host material for highly nonlinear HFs. Although their intrinsic material nonlinearity (and indeed linear refractive index) is lower than chalcogenide and heavy metal oxide glasses [43], they offer higher thermal and crystallization stability and less steep viscosity-temperature-curves, while exhibiting low softening temperatures [44]. Among commercially available lead silicate glasses, the Schott glass SF57 exhibits the highest nonlinearity. The softening temperature of SF57 is 520 °C [45], the nonlinear refractive index of this glass was measured to be 4.1×10^{-19} m^2/W at1060 nm [40], and the linear refractive index of SF57 is ≈1.8 at 1550 nm [45]. The zero-dispersion wavelength for the glass is 1970 nm and the material dispersion is strongly normal at 1550 nm. Compared with chalcogenide and heavy metal oxide glasses (n~2.4 at 1550 nm), the significantly lower refractive index of SF57 should enable more efficient integration with conventional fiber systems.

Bismuth: Bismuth oxide-based glass is an attractive novel material for nonlinear devices. It shows high nonlinearity but without containing toxic elements such as Pb, As, Se, Te [46]. Moreover, the bismuth-based glass exhibits good mechanical, chemical and thermal stability, which allows easy fiber fabrication process. A nonlinear fiber [47] and a short Er-doped fiber amplifier with broadband emission [48] have been developed from this glass. In addition, bismuth-oxide-based fibers can be fusion-spliced to silica a fiber [49], which offers easy integration to silica-based networks. The fibers review in this paper were made from bismuth-oxide-based glass developed at Asahi Glass Company. Due to the high bismuth content, the glass exhibits a high linear and nonlinear refractive index of $n = 2.02$ and $n_2 = 3.2\times10^{-19}$ m^2/W at 1550 nm, respectively [47], and this glass has a softening temperature of 550 °C.

Tellurite: Tellurite glasses, like lead silicate and bismuth glasses, offer high refractive index and high optical nonlinearity ($n_2=2.5\times10^{-19}$ m^2/W) [29]. In addition, tellurite glass has good infrared transmittance and has a low phonon energy relative to other oxide glasses [50]. Furthermore, tellurite glasses are more stable than fluoride glasses, have higher rare-earth solubilities than chalcogenide glasses [50] and have an order of magnitude larger Raman gain peak than fused silica [51]. Tellurite glasses have a low softening temperature around 350 °C [50].

High nonlinearity microstructured fibers: Key results to date

- SF57 holey fiber with an effective nonlinearity coefficient of $\gamma = 640W^{-1}km^{-1}$ (loss ≈ 2.6dB/m at 1550nm) [31]
- Anomalous dispersion at 1550nm and Raman soliton generation in SF57 holey fiber [31]
- Observation of the soliton self frequency shift and pulse compression in SF57 holey fiber [31]
- Tellurite holey fiber with $\gamma = 48W^{-1}km^{-1}$ (loss ≈5dB/m at 1550nm) [29]
- Observation of first and second order Stokes stimulated Raman scattering in 1m of tellurite holey fiber [29]
- Bismuth-oxide-based glass holey fiber with $\gamma = 460$ W^{-1} km^{-1} [33]
- Demonstration of splicing of bismuth holey fibers to conventional fibers [34]
- Demonstration of supercontinuum generation in SF6 holey fiber [28]

Discussion:

Both the effective fiber nonlinearity and effective fiber length determine the performance of a nonlinear device (the effective length depends on the propagation loss). Small-core high-NA holey fibers based on bismuth-oxide and lead-silicate glasses have clearly higher fiber nonlinearity but also higher propagation loss compared with highly nonlinear silica holey fibers. However, for short devices using ≤ 1 m fiber length, fiber losses of ≤ 2 dB/m can be readily tolerated. For ≤ 2 dB/m, the effective fiber length is $\geq 80\%$ of the real fiber length of ≤ 1 m, whereas the nonlinearity of non-silica fibers can be up to 10 times higher than that of silica holey fibers. In other words, in compact devices using ≤ 1 m fiber length with ≤ 2 dB/m loss, the increase in fiber nonlinearity obtained by using nonlinear glass compositions clearly outweighs the decrease of the effective fiber length due to higher propagation losses. Thus, provided that relatively low loss fibers can be produced, highly nonlinear compound glass holey fibers can exhibit a better nonlinear performance (lower power or shorter fiber length requirement) compared with nonlinear silica holey fibers.

Most nonlinear/high-index glasses have a high normal material dispersion at 1550 nm, which tends to dominate the overall dispersion of fibers with a conventional solid cladding structure. However, for many nonlinear device applications, anomalous or near-zero dispersion is required. Fortunately, the cladding geometry significantly affects the waveguide dispersion, allowing the highly normal material dispersion to be overcome. Indeed, for example, lead silicate holey fibers showing anomalous dispersion at 1550 nm have been demonstrated [32]. The fact that fiber dispersion is anomalous also enables us to exploit soliton effects [31].

To make practical fiber-connectorized devices, splicing of compound glass holey fibers to silica fibers is important. Recently, bismuth-oxide based holey fiber with $A_{eff} \sim 2.8\ \mu m^2$ has been spliced to a silica SMF28 patchcord using a conventional mechanical cleaver and fusion splicer [34]. To reduce the overall mode-mismatch loss, two intermediate buffer stages were used. The bismuth glass holey fiber itself was spliced to a silica fiber with $A_{eff} \approx 14 \mu m^2$. Due to the much lower melting temperature of the bismuth glass relative to silica glass, very small values for the fusion time and current were used. The splices achieved were mechanically strong, especially with respect to applied strain in the axial direction (Figure 3a). Although the total splicing losses achieved to date are still quite high (5.8 dB) – they can largely be accounted for by individual mode-mismatches at the various buffer fiber interfaces (minimum 3.8 dB, without taking into account the mismatch in the mode shape) and an additional 0.1 dB due to Fresnel reflection at the bismuth glass HF / silica fiber interface. There is considerable scope for reducing the Fresnel reflections using accurately controlled angle cleaves at the bismuth glass /silica fiber splice. The introduction of an additional silica HF based buffer stage should help to considerably further reduce the mode-mismatch. Splicing of bismuth glass holey fiber to silica fiber has resulted in two benefits in the fiber performance. One is the reduction of coupling losses between the two fibers by 0.9 dB relative to butt coupling. The other is the achievement of robust single-mode guidance in the bismuth glass holey fiber at 1550 nm, although the fiber can support more than one mode in case of free-space coupling. An IR image of the far end of the connectorized holey fiber, when laser pulses at 1558 nm were launched into the fiber, showed that only the fundamental mode was excited (Figure 3b). (The image also demonstrates that the fundamental mode of the holey fiber has a triangular mode shape, in agreement with the predicted mode profile [33].)

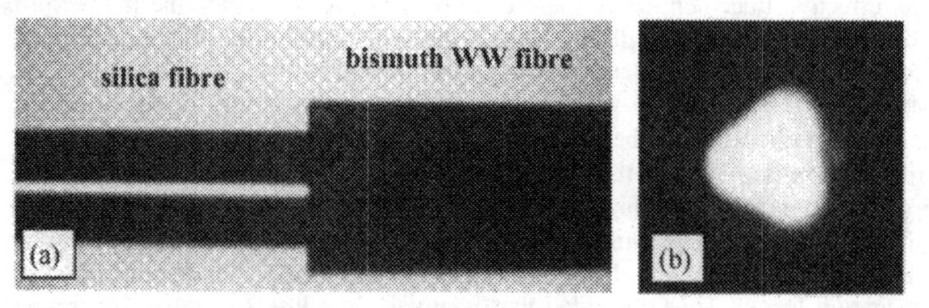

Figure 3 (a) Microscope image of bismuth glass holey fiber to silica fiber splice, (b) IR image of the near-field pattern of the connectorized holey fiber.

NEW TRANSMISSION FIBERS

A range of non-silica glasses can exhibit properties such as transparency in the mid-IR region and high solubility of rare-earth ions that are not available in silica glass and hence non-silica HFs are of particular promise for applications in mid-IR region and active devices. One challenge for high power applications in compound glass-based fibers is the onset of intensity-dependent nonlinear effects. The use of single-mode HF large-mode-area (LMA) designs [8] is one means of minimizing such nonlinear effects. Since the nonlinear refractive index (n_2) of non-silica glasses is typically in the range $(1\text{-}50)\times10^{-19}m^2/W$, higher than that of silica glass (2.7×10^{-20} m^2/W) by 1-2 orders of magnitude, very large-mode-areas ($\gg 100\mu m^2$) are required to reduce the effective nonlinearity γ to below $60W^{-1}km^{-1}$. In practice, it is challenging to fabricate compound glass LMA HFs with the complex microstructured cladding structures required to provide low NA guidance (i.e. a large number of relatively small air holes). Here we review recent work on the development of large mode area single mode lead silicate holey fiber.

A holey fiber with a mode area of $40\mu m^2$ at 800nm was fabricated from the Schott lead silicate glass SF6 using the conventional capillary-stacking technique. Previously, this technique has principally been used for the fabrication of silica HFs. The cross-sectional microstructure of this HF is shown in Figure 4 (a) & (b), where a four-ring microstructured cladding can be seen. The hole-to-hole spacing Λ in the hexagonally arrayed holey cladding was measured to be 4.3µm. In addition to the principal holes with an inner-diameter (ID) of 2.7µm (d_1), six smaller holes with an ID of 0.3µm (d_2) are periodically distributed in the second ring surrounding the solid core. For the holes near the solid core, i.e., the holes with sizes d_1 and d_2, the ratio of hole-diameter to the hole-to-hole spacing Λ is $d_1/\Lambda=0.63$, and $d_2/\Lambda=0.07$, respectively. The 5.5µm (d_3) holes only weakly influence the optical guidance properties of this HF as they are located far from the core. Robust single mode guidance, from at least 700 to 800nm was observed in this HF. The effective mode area of the fundamental mode $A_{eff}=40\mu m^2$ was extracted from the measured mode intensity profile. Although this value of mode area is still not large enough to avoid intensity-dependent nonlinear effects in this SF6-based HF, this work demonstrates that the capillary-stacking technique can be successfully used to fabricate soft-glass HFs with a complex holey cladding. We anticipate that high-index glass based single-mode HF with very large mode areas (>>100μm^2) should be possible using this technique in the future.

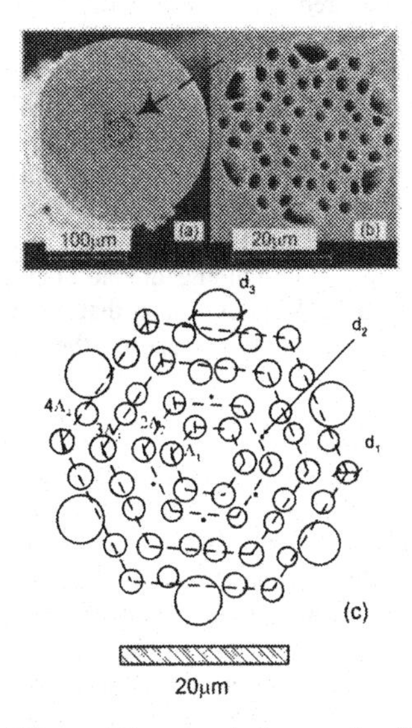

Figure 4 (Top) SEM (Scanning Electron Microscope) photographs of SF6 glass holey fiber. (Bottom) Schematic of cladding design ($\Lambda = (\Lambda_1 \approx \Lambda_2 \approx \Lambda_3 \approx \Lambda_4) = 4.3 \pm 0.2$ μm).

SOLID MICROSTRUCTURED FIBERS

Motivation

Research to date on silica microstructured fibers has shown that the combination of wavelength-scale features and a large refractive index contrast is a powerful means of obtaining fibers with a broad range of useful optical properties. However, there are some practical drawbacks associated with the use of air/glass fibers. When compared with solid fibers, air/glass microstructured fibers are challenging to splice, polish, taper, and, when the cross-section is largely comprised of air, they can be fragile. In addition, it can be challenging to fabricate kilometer-scale holey fibers with identical and controllable cladding configurations. This is because the transverse profile of a drawn microstructured fiber is sensitive to the effect of pressure inside the holes, surface tension at the air/glass boundaries and temperature gradients present during the fiber drawing process. To prevent the collapse of the holey microstructure in HFs, the holes within the preform are often sealed, and consequently the air pressure inside the structure changes during fiber drawing. Hence the precise details of the final fibre microstructure are typically time-dependent as well as dimension-dependent.

Note that the optical characteristics of microstructured fibers can be sensitive to the cladding configuration and even minor changes in the microstructure can cause noticeable deviations in properties such as dispersion.

The development of microstructured fibers with solid microstructured claddings promises to eliminate these drawbacks. Two different implementations of this concept have recently been realized. The first approach is to replace the air holes in the transverse structure of an air/glass microstructured preform with low index glass regions to produce an index-guiding solid holey fiber [36]. Another approach is to form a fiber in which the cladding is defined by a series of thin nested layers of glass (a Bragg-type fiber cladding) to form an air-guiding fiber [37]. In both approaches, the basic requirement is the identification of materials which are thermally and chemically matched in order that they can be drawn into optical fiber and which provide a sufficient index contrast to allow light to be confined either by index-guiding or photonic bandgap effects without requiring unfeasibly large numbers of cladding features. Note that while this approach may have some practical advantages, as suggested above, it clearly restricts the range of materials that can be used.

Solid Holey Fiber

Here we review the recent development of all-solid index-guiding holey (SOHO) fiber based on two thermally-matched silicate glasses with a high index contrast (as described in Reference [36]).

Fabrication: A borosilicate glass containing lead-oxide (PbO >30mol.%) with a refractive index of n = 1.76 at 1.55μm, was selected as the background material (labeled B1 thereafter) for this SOHO fiber. Another silicate glass containing high alkali-oxide (Na_2O, K_2O etc) with index n = 1.53 at 1.55 μm was selected as the material to fill in the holes (H1). These glasses were selected because of their mechanical, rheological, thermo-dynamic and chemical compatibility. Rods of H1 glass were inserted into B1 tubes, both of which were drilled from bulk glass samples. These rod/tube structures were caned on a fiber drawing tower and stacked around a core rod of B1 glass within a B1 glass jacket tube using the capillary-stacking technique that is used to produce silica HFs. This structured preform was then drawn using a two-stage drawing procedure to produce two fibers: one with an outer diameter (OD) of 440μm and the other with an OD of 220μm.

Configuration of solid-microstructured cladding: Figure 5 shows cross-sectional profiles of the 220μm OD fiber taken using a Scanning Electron Microscope (SEM). From Figures 5 (a) & (b), it can be seen that this SOHO fiber is indeed an all-solid fiber without any observable air holes. However when the accelerating voltage (EHT) of the SEM is increased from 2.72kV to 22.00 kV, the microstructure in the cladding region becomes apparent. This occurs because the backscattering coefficient of the primary electrons from the SEM is a monotonically increasing function of the atomic number [52]. As a result, glasses with low average atomic number and consequently with the low density exhibit relatively lower brightness than materials with high density. Thus, as shown in Figures 5(c) and (d), the large density difference between B1 and H1 glasses (5.2 g/cm^3 and 2.9 g/cm^3 respectively) it is possible to distinguish the distribution of each material within the fiber profile.

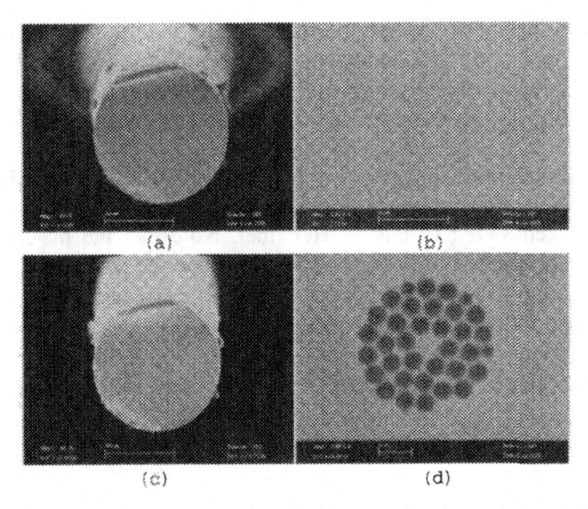

Figure 5: SEMs of 220 μm diameter SOHO fiber by adjusting accelerating voltage (EHT) (a) whole view, EHT = 2.72 kV, (b) zoomed center view, EHT = 2.72 kV; (c) whole view, EHT = 22.00 kV, (d) zoomed center view, EHT = 22.00 kV

Typically when air/glass holey fibers are drawn, the holes become non-circular due to surface tension effects, which modify the geometry of the cladding. Hence the details of the resulting cross-sectional profiles depend on the draw-down ratio and the drawing tension, and this becomes particularly apparent when small-scale features are required. Since the optical properties of such fibers can be sensitive to the cladding configuration, this effect can make it difficult to predetermine the preform required for a given fiber specification. Figure 6 compares the all-solid claddings of the 1mm cane, 440 μm fiber and 220 μm fiber. First, note that all the low index (black) regions retain their circularity, regardless of the draw-down ratio. Second, the d/Λ ratio (d: average diameter of the low index regions, Λ: the pitch (center to center spacing) of these regions) is ~0.81 regardless of the fiber OD. SOHO fibers thus provide a practical way of avoiding structural deformations during fiber drawing.

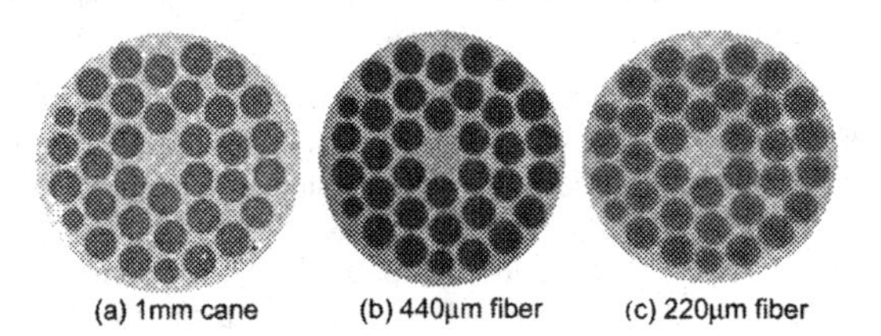

Figure 6. SEMs of microstructured cladding in (a) 1mm cane inserted into jacket tube before drawing, (b) 440 μm OD fiber with Λ = 4μm and (c) 220μm OD fiber with Λ = 2μm. (EHT = 22 kV)

Fiber attenuation: The index contrast of B1/H1 SOHO fiber is $\Delta n = 0.23$ at 1.55 μm, less than that in silica/air HF ($\Delta n = 0.44$ at 1.55μm), which might at first sight be expected to result in poorer mode confinement. However, the refractive index of the background glass, H1 (n=1.76 at 1.55μm) is higher than that of silica (n=1.44 at 1.55μm) which might be expected to lead to improved mode confinement. To resolve which factor dominates, Figure 7(a) shows the calculated confinement losses of the fundamental mode in a range of B1/H1 based SOHO fibers. These predictions were made with the multipole technique [53], which can model the properties of fibers with circular holes, so it is particularly well suited to this SOHO fiber. Note that the material absorption and the fabrication defects have been ignored and it is assumed that the holes lie on a hexagonal lattice. It can be seen that when $\Lambda = 4$μm, the confinement losses decrease rapidly to less than 0.001 dB/m by (1) increasing d/Λ ratio from 0.5 to 0.8 or more, or (2) by increasing the number of the rings from 2 to 4 or more with a fixed d/Λ = 0.5. Figure 7(a) indicates that is possible to design B1/H1 SOHO fibers (when $\Lambda = 4$μm) with negligible confinement loss provided that the d/Λ ratio is sufficiently large (0.8 or larger), as is the case for our 440 μm OD SOHO fiber.

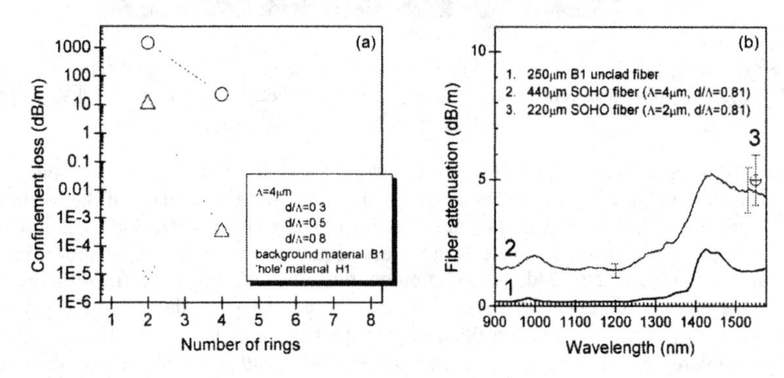

Figure 7(a) Calculated confinement losses of B1/H1 based SOHO fiber as a function of the number of hexagonal packed rings and their diameter to spacing ratio d/Λ with Λ=4μm at 1.55μm; (b) measured propagation attenuation of (1) unclad 250 μm B1 fiber, (2) 440 μm B1/H1 SOHO fiber, (3) 220 μm B1/H1 SOHO fiber. Note that (3) is a spot measurement at 1550nm. Measurement errors are plotted.

Figure 7(b) shows the measured fiber attenuation for the case of a 250 μm unclad/unstructured B1 fiber, 440 μm B1/H1 SOHO fiber with $\Lambda = 4$μm and d/Λ = 0.81, and 220μm B1/H1 SOHO fiber with Λ=2μm and d/Λ = 0.81. It can be seen that the attenuation of both fibers for which $\Lambda = 4$ μm and 2 μm is ~5 dB/m at 1.55 μm, implying that as predicted, the confinement loss of the fundamental mode is negligible for these fibers and the loss is dominated by the material absorption and other factors related to the fiber fabrication process. In detail, due to the overtone of the fundamental vibration of hydrogen bonding in B1 glass, the attenuation of all the fibers increases around 1.5 μm, while between 1.0-1.2 μm the fibers show minimum losses of ~1.5 dB/m. Additionally, due to the similar bonding strengths of Si-O bonds and Pb-O bonds, near fiber drawing temperatures, the SiO_2-PbO glass system can separate into SiO_2-rich regions and PbO-rich regions at the sizes of micron or sub-micron scale. Such phase-separation can lead to compositional variations (and thus the index variations) within the fiber profile. Impurities in the bulk glasses are another contribution to the total fiber loss. By using high purity raw materials and melting the glasses in a dry atmosphere [54], it should ultimately be possible to reduce the total loss of this SOHO fiber at 1.55 μm to below the 1 dB/m level, thus opening up many practical applications for this new fiber type.

Fiber nonlinearity: One of the most attractive applications for holey fibers made from high index glasses is in nonlinear devices. In order to estimate the effective fiber nonlinearity that can be achieved in this material system, Figure 8 shows calculations of the effective mode area and corresponding fiber nonlinearity γ for a rod of material B1 surrounded by a uniform non-structured cladding of material H1. This simplified geometry represents the fundamental limit in effective mode area/nonlinearity that can be achieved in a fiber made from these two materials. The smallest mode area (and hence highest γ) occurs when the core is ~1 μm in diameter. Corresponding results for a silica rod suspended in air are also shown in Figure 8. Even though the index contrast between silica/air leads to a similar mode area as the combination B1/H 1, the significantly larger material nonlinearity (n_2) of material B1 results in a dramatic improvement in the nonlinearity. Hence while the maximum nonlinearity that can be achieved in a silica/air holey fiber is ~60 $W^{-1}km^{-1}$, more than 500 $W^{-1}km^{-1}$ should ultimately be possible in a B1/H1 SOHO fiber. The Boskovic method [55] was applied to measure the effective nonlinearity of the 220μm SOHO fiber with $\Lambda = 2\mu m$ and $d/\Lambda = 0.81$. Using high power dual frequency beat signals, the effective fiber nonlinearity γ was deduced from the nonlinear phase shift F_{SPM}, ($F_{SPM} = 2\gamma LP$, where L is the effective fiber length and P the signal power), due to the propagation in the fiber to be 230 $W^{-1}km^{-1}$, which is ~200 times higher than that of standard single mode silica fiber and matches the modeled result well.

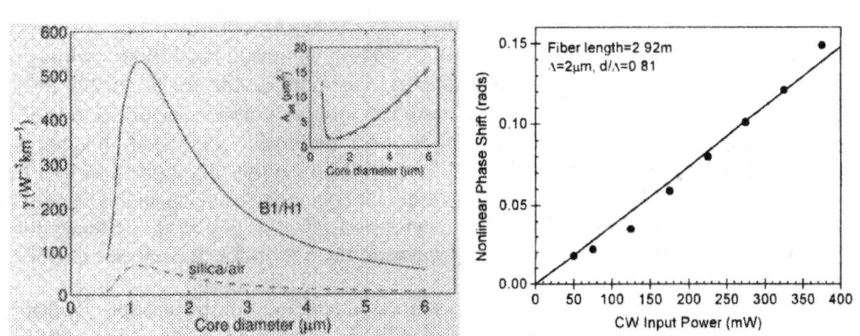

Figure 8. Effective nonlinearity of SOHO fiber (left: calculated for a range of simple step-index fiber designs, right: measured relationship between nonlinear phase shift and the input laser power at 1.55μm)

Group velocity dispersion (GVD): One of the most important applications of HFs is highly nonlinear fibers. However, the high dispersion and dispersion slope which are characteristic of many

HF designs limits the useful spectral bandwidth of the fibers. In silica HFs, low and flat dispersion can be obtained when d/Λ on the order of 0.25-0.3 with $\Lambda = 2.3\mu m$ [56,57]. Hence for the low/flat dispersion air/silica HF designs identified thus far, these GVD properties are achieved at the cost of a considerable reduction in nonlinearity relative to small core large air-fraction HFs. In contrast, the predictions in Figure 9 suggest that zero GVD in the 1.5-1.6μm wavelength range can be achieved in realistic B1/H1 SOHO fiber designs with large d/Λ and small Λ, which implies that high nonlinearity and low dispersion can be achieved simultaneously in solid-microstructured fibers.

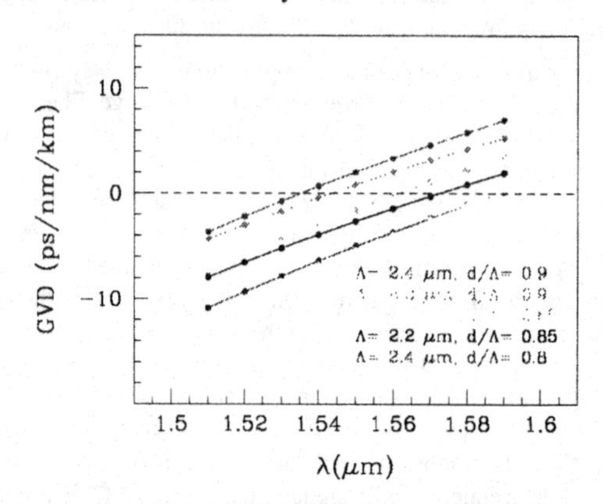

Figure 9. GVD predictions for a range of B1/H1 SOHO fibers made using a full-vector implementation of the orthogonal function method [56]. The material dispersions of B1 and H1 have been included ab initio.

Hollow core fiber with multilayered cladding

Here a brief overview of another class of solid cladding microstructured fiber: hollow core fiber with a solid multilayered photonic bandgap cladding. For more details see Reference [37]. In this work, which was conducted at MIT, a fiber designed for the transmission of 10.6μm CO_2 laser emission was producing using a combination of polymer and chalcogenide glass materials that are not themselves transparent at the operating wavelength. The periodically structured solid cladding effectively acts as a mirror for light propagating within the photonic bandgap, and light is confined to the hollow core of the fiber. The relatively low fiber losses achieved (<1dB/m) are made possible by the short penetration depths of electromagnetic waves in the high-refractive-index-contrast photonic crystal structure.

Fabrication: To achieve high index contrast in the layered portion of the fiber, a chalcogenide glass with a refractive index of 2.8 (arsenic triselenide) was combined with a high glass-transition temperature thermoplastic polymer with a refractive index of 1.55, poly(ether sulphone). These materials are thermally compatible and can be drawn into good quality layered structures. The glass layers were thermally evaporated onto a polymer film, and the coated film was rolled into a hollow multilayer tube fiber preform. This hollow macroscopic pre-form was drawn down into fiber with submicron layer thicknesses. Transmission spectra were used to confirm that the positions of the measured transmission peaks agree with the calculated bandgaps, which confirms that light is guided in the hollow core by the photonic bandgap mechanism.

Power delivery: Using a fiber with a fundamental photonic bandgap centered near 10.6μm and a hollow core with a diameter of 700μm, CO_2 laser emission at 10.6μm was successfully transmitted. This demonstrates the substantial reduction of fiber transmission loss relative to the intrinsic loss of the constituent materials. No damage to the fibers was reported when a laser power density of $\approx 300 W/cm^2$ was coupled into the hollow fiber core. These results indicate the feasibility of using hollow multilayer photonic bandgap fibers as a transmission medium for high-power laser light in the infrared.

CONCLUSION

The synergy of novel compound glass materials and microstructured fiber technology promises a broad range of new and potentially useful optical fibers. The low softening temperatures characteristic of compound glasses allow a broad range of fabrication techniques to be exploited for the production of macroscopic fiber preforms. The technique that has been developed most extensively to date is extrusion, which has proved to be a flexible approach to realizing fibers from relatively small quantities of bulk glass. Using conventional solid fiber designs, it is in general necessary to identify two compatible glasses for the production of low loss optical fibers. Note that since microstructured fibers can be made from a single material, this technology is a powerful way of producing fibers from glasses with useful optical properties without requiring that compatible cladding materials be found.

A range of theoretical techniques has been developed in recent years to model the optical properties of silica microstructured fibers. These techniques have been applied to several non-silica microstructured fiber designs (both index-guiding and photonic bandgap). The relatively large index contrast typically found in microstructured fibers made from high index glasses makes fiber design more challenging, and in general means that it is more important to use full vectorial methods for these fibers. Numerical predictions have been used to usefully complement ongoing fabrication and experimental work for a range of index-guiding fiber designs (see for example [32]).

Using preform extrusion techniques, fibers with high effective nonlinearity have been made from a range of high-n_2 non-silica glasses. The largest reported value of effective fiber nonlinearity (γ) in a holey fiber has been realized using lead silicate glass, where $\gamma \sim 640 W^{-1}km^{-1}$ has been achieved. Theoretical work indicates that values of γ of a few thousand $W^{-1}km^{-1}$ should ultimately be possible. The best fibers of this type to date have propagation losses of approximately 3 dB/m [32]. Given that this fiber type has only emerged relatively recently, it is anticipated it should be possible to reduce the losses the 1 dB/m level or below for some high n_2 glasses. Such an advance should enable this new fiber type to compete (in terms of device figure of merits) with more conventional fibers for nonlinear fiber devices. Work on improving on the design of such fibers as well as developing techniques that will allow more efficient coupling to conventional systems is currently under way.

The large index contrast possible in non-silica microstructured fibers is promising for the development of new fibers based on photonic bandgap effects. It has been demonstrated that increasing the refractive index contrast beyond that available in air/silica does not necessarily broaden the photonic bandgaps that are available [58,59]. Unlike air/silica bandgap fibers, in which the photonic bandgaps widen monotonically as a function of increasing air filling fraction, the optimum air-filling fraction for high index glasses (i.e. $n>2$) is approximately 60% (this is referred to as type-II photonic bandgap guidance). From a practical viewpoint, this is fortuitous, since non-silica glasses tend to be relatively fragile, and it would be challenging to produce the extremely large air filling fraction designs that are typically made in air/silica structures [26]. One particularly promising application of this new fiber type will be the development of new air-guiding fibers for broadband high power IR transmission [59].

It has been demonstrated that solid microstructured fibers can be produced using a relatively small number of cladding features. Large effective nonlinearity can be combined with relatively low and flat group velocity dispersion between 1.5-1.6μm in this class of fiber. Additionally, it can be expected that the use of a solid fiber structure will lead to a number of practical advantages relative to air/glass HFs. For example, edge polishing, angle polishing and splicing should all be more straightforward in SOHO fibers [36]. Light can also be guided in air in fibers with a solid microstructured cladding [37]. In these fibers, chalcogenide/polymer cladding layers act as an interior omnidirectional dielectric mirror confining light to the hollow core via photonic bandgap effects. The transmission windows of these fibers are determined by the layer dimensions and can be scaled from 0.75 to 10.6μm in wavelength. The transmission losses are orders of magnitude lower than those of the materials used to make the fiber cladding, thus demonstrating that low attenuation can be achieved through structural design rather than high-transparency material selection.

Non-silica microstructured fibers have the potential to dramatically broaden the range of optical properties that can be offered in fiber form. For example, the combination of the high nonlinearity provided by compound glasses and the flexibility offered by microstructured fiber technology for engineering the fiber parameters should lead to the development of truly practical, low power compact nonlinear fiber devices. Solid holey fibers promise both a range of practical handling benefits and access to new optical regimes. Finally, air-core high-index photonic bandgap fibers are an attractive means to high-power transmission and new transmission windows.

REFERENCES

1. J.C. Knight, T.A. Birks, P.S.J. Russell, D.M. Atkin, "All-silica single-mode optical fiber with photonic crystal cladding", Opt. Lett. **21**, pp. 484-485, 1996
2. T.M. Monro, D.J. Richardson, "Holey optical fibres: Fundamental properties and device applications", C. R. Phys. **4**, pp. 175-186, 2003
3. P. Russell, "Photonic crystal fibers", Science **299**, pp. 358-362, 2003
4. R.F. Cregan, B.J. Mangan, J.C. Knight, T.A. Birks, P.S. J. Russell, P.J. Roberts, D.C. Allan, "Single-mode photonic band gap guidance of light in air", Nature **285**, pp.1537-1539, 1999
5. T.M. Monro, P.J. Bennett, N.G.R. Broderick, D.J. Richardson, "Holey fibers with random cladding distributions", Opt. Lett. **25**, pp. 206-208, 2000
6. T.A. Birks, J.C. Knight, P.S.J. Russell, "Endlessly single-mode photonic crystal fiber", Opt. Lett. **22**, pp. 961-963, 1997
7. J.C. Knight, J. Arriaga, T.A. Birks, A. Ortigosa-Blanch, W.J. Wadsworth, P.S.J. Russell, "Anomalous dispersion in photonic crystal fiber", IEEE Photon. Technol. Lett. **12**, pp. 807-809, 2000

8. J.C. Knight, T.A. Birks, R.F. Cregan, P.S.J. Russell, J.-P. de Sandro, "Large mode area photonic crystal fibre", Electron. Lett. **34**, pp. 1347-1348, 1998
9. B. Zsigri, C. Peucheret, M.D. Nielsen, P. Jeppesen, "Transmission over 5.6km large effective area and low-loss (1.7dB/km) photonic crystal fibre", Electron. Lett. **39**, pp. 796-798, 2003
10. P. Petropoulos, M. Monro, W. Belardi, K. Furusawa, J.H. Lee, D.J. Richardson, "2R-regenerative all-optical switch based on a highly nonlinear holey fiber", Opt. Lett. **26**, pp. 1233-1235, 2001
11. A.I. Siahlo, L.K. Oxenlowe, K.S. Berg, A.T. Clausen, P.A. Andersen, C. Peucheret, A. Tersigni, P. Jeppesen, K.P. Hansen, J.R. Folkenberg, "A high-speed demultiplexer based on a nonlinear optical loop mirror with a photonic crystal fiber", IEEE Photon. Technol. Lett. **15**, pp. 1147-1149, 2003
12. J.H. Lee, P.C. Teh, W. Belardi, M. Ibsen, T.M. Monro, D.J. Richardson, "A tunable WDM wavelength converter based on cross-phase modulation effects in normal dispersion holey fiber", IEEE Photon. Technol. Lett. **15**, pp. 437-439, 2003
13. J.E. Sharping, M. Fiorentino, P. Kumar, R.S. Windeler, "Optical parametric oscillator based on four-wave mixing in microstructure fiber", Opt. Lett. **27**, pp. 1675-1677, 2002
14. C. Peucheret, B. Zsigri, P.A. Andersen, K.S. Berg, A. Tersigni, P. Jeppesen, K.P. Hansen, M D. Nielsen, "40Gbit/s transmission over photonic crystal fibre using mid-span spectral inversion in highly nonlinear photonic crystal fibre", Electron. Lett. **39**, pp. 919-921, 2003
15. J.K. Ranka, R.S. Windeler, A.J. Stentz, "Visible continuum generation in air-silica microstructure optical fibers with anomalous dispersion at 800nm", Opt. Lett. **25**, pp. 25-27, 2000
16. P.A. Champert, S.V. Popov, J.R. Taylor, "Generation of multiwatt, broadband continua in holey fibers", Opt. Lett. **27**, pp. 122-124, 2002
17. R. Holzwarth, T. Udem, T.W. Hansch, J.C. Knight, W.J. Wadsworth, P.S.J. Russell, "Optical frequency synthesizer for precision spectroscopy", Phys. Rev. Lett. **85**, pp. 2264-2267, 2000
18. I. Hartl, X.D. Li, C. Chudoba, R.K. Ghanta, T.H. Ko, J.G. Fujimoto, J.K. Ranka, R.S. Windeler, "Ultrahigh-resolution optical coherence tomography using continuum generation in an air-silica microstructure fiber", Opt. Lett. **26**, pp. 608-610, 2001
19. W.J. Wadsworth, J.C. Knight, A. Ortigosa-Blanch, J. Arriaga, E. Silvestre, P.S.J. Russell, "Soliton effects in photonic crystal fibres at 850nm", Electron. Lett. **36**, pp. 53-55, 2000
20. B.R. Washburn, S.E. Ralph, P.A. Lacourt, J.M. Dudley, W.T. Rhodes, R.S. Windeler, S. Coen, "Tunable near-infrared femtosecond soliton generation in photonic crystal fibres", Electron. Lett. **37**, pp. 1510-1512, 2001
21. Z. Yusoff, J.H. Lee, W. Belardi, T.M. Monro, P.C. Teh, D.J. Richardson, "Raman effects in a highly nonlinear holey fiber: amplification and modulation", Opt. Lett. **27**, pp. 424-426, 2002
22. T. Hasegawa, E. Sasaoka, M. Onishi, M. Nishimura, Y. Tsuji, M. Koshiba, "Hole-assisted lightguide fiber for large anomalous dispersion and low optical loss", Opt. Express **9**, pp. 681-686, 2001
23. J.K. Sahu, C.C. Renaud, K. Furusawa, R. Selvas, J.A. Alvarez-Chavez, D.J. Richardson, J. Nilsson, "Jacketed air-clad cladding pumped ytterbium-doped fibre laser with wide tuning range", Electron. Lett. **37**, pp. 1116-1117, 2001

24. J.H.V. Price, K. Furusawa, T.M. Monro, L. Lefort, D.J. Richardson, "Tunable, femtosecond pulse source operating in the range 1.06-1.33μm based on an Yb^{3+}-doped holey fiber amplifier", J. Opt. Soc. Amer. B **19**, pp. 1286-1294, 2002
25. K. Tajima, J. Zhou, K. Kurokawa, K. Nakajima, "Low water peak photonic crystal fibres", Proc. European Conference on Optical Communication (ECOC'2003), Rimini, Italy, 2003, postdeadline paper Th4.1.6
26. Brian Mangan, Lance Farr, Allen Langford, P. John Roberts, David P. Williams, Francois Couny, Matthew Lawman, Matthew Mason, Sam Coupland, Randolf Flea, Hendrik Sabert, Tim A. Birks, Jonathan C. Knight, Philip St. J. Russell, "Low loss (1.7 dB/km) hollow core photonic bandgap fiber", PDP24 OFC'2004 (Anaheim).
27. K.M. Kiang, K. Frampton, T.M. Monro, R. Moore, J. Trucknott, D.W. Hewak, D.J. Richardson, "Extruded singlemode non-silica glass holey optical fibres", Electron. Lett. **38**, pp. 546-547, 2002
28. V.V.R.K. Kumar, A.K. George, W.H. Reeves, J.C. Knight, P.S.J. Russell, "Extruded soft glass photonic crystal fiber for ultrabroadband supercontinuum generation", Opt. Express **10**, pp. 1520-1525, 2002
29. V.V.R.K. Kumar, A.K. George, J.C. Knight, P.S.J. Russell, "Tellurite photonic crystal fiber", Opt. Express **11**, pp. 2641-2645, 2003
30. T.M. Monro, K.M. Kiang, J.H. Lee, K. Frampton, Z. Yusoff, R. Moore, J. Trucknott, D.W. Hewak, H.N. Rutt, D.J. Richardson, "High nonlinearity extruded single-mode holey optical fibers", Proc. Optical Fiber Communications Conference (OFC'2002), Anaheim, California, 2002, postdeadline paper FA1-1
31. P. Petropoulos, T.M. Monro, H. Ebendorff-Heidepriem, K. Frampton, R.C. Moore, H.N. Rutt, D.J. Richardson, "Soliton-self-frequency-shift effects and pulse compression in anomalously dispersive high nonlinearity lead silicate holey fiber", Proc. Optical Fiber Communications Cnference (OFC'2003), Atlanta, Georgia, 2003, postdeadline paper PD03
32. P. Petropoulos, H. Ebendorff-Heidepriem, V. Finazzi, R.C. Moore, K. Frampton, D.J. Richardson, T.M. Monro, "Highly nonlinear and anomalously dispersive lead silicate glass holey fibers", Opt. Express **11**, pp. 3568-3573, 2003
33. H. Ebendorff-Heidepriem, P. Petropoulos, V. Finazzi, K. Frampton, R.C. Moore, D.J. Richardson, T.M. Monro, "Highly nonlinear bismuth-oxide-based glass holey fiber, Proc. Optical Fiber Communications Conference (OFC'2004), Los Angeles, California, 2004, paper ThA4
34. P. Petropoulos, H. Ebendorff-Heidepriem, T. Kogure, K. Furusawa, V. Finazzi, T.M. Monro, D.J. Richardson, "A spliced and connectorized highly nonlinear and anomalously dispersive bismuth-oxide glass holey fiber", Proc. Conference on Lasers and Electro-Optics (CLEO'2004), San Francisco, California, 2004, paper CTuD
35. M.A. van Eijkelenborg, M.C.J. Lange, A. Argyros, J. Zagari, S. Manos, N.A. Issa, S. Fleming, R.C. McPhedran, C.M. de Sterke, N.A.P. Nicorovici, "Microstructured polymer optical fibre", Opt. Express **9**, pp. 319-327, 2001
36. X. Feng, T.M. Monro, P. Petropoulos, V. Finazzi, D.W. Hewak, "Solid microstructured optical fiber", Opt. Express **11**, pp. 2225-2230, 2003

37. Burak Temelkuran, Shandon D. Hart, Gilles Benoit, John D. Joannopoulos & Yoel Fink, "Wavelength-scalable hollow optical fibres with large photonic bandgaps for CO_2 laser transmission", Nature **420**, pp. 650-653, 2002
38. G. P. Agrawal, Nonlinear Fiber Optics (Academic Press, Boston, 2001).
39. J. H. Lee, W. Belardi, K.Furusawa, P. Petropoulos, Z.Yusoff, T.M. Monro, and D.J. Richardson, "Four-wave mixing based10-Gb/s tunable wavelength conversion using a holey fiber with a high SBS threshold," IEEE Photon. Technol. Lett. **15**, 440-442 (2003).
40. S.R. Fribergand P.W. Smith, "Nonlinear Optical-Glasses for Ultrafast Optical Switches," IEEEJ. Quantum Electron. 23, 2089-2094 (1987).
41. R.E. Slusher, J.S. Sanghera, L.B. Shaw, I.D. Aggarwal, "Nonlinear optical properties of As-Se fiber", Proceedings of OSA Topical meeting on Nonlinear Guided Waves and their Applications, Stresa, Italy, 1-4 Sep 2003
42. L. Poladian, N. A. Issa, and T. M. Monro, "Fourier decomposition algorithm for leaky modes of fibres with arbitrary geometry," Opt. Express **10**, pp. 449-454, 2002
43. E. M. Vogel, M. J. Weber, and D. M. Krol, "Nonlinear Optical Phenomena in Glass," Phys. Chem. Glasses **32**, pp. 231-254, 1991
44. S. Fujino, H. Ijiri, F. Shimizu, and K. Morinaga, "Measurement of viscosity of multi-component glasses in the wide range for fiber drawing," J. Jpn. Inst. Met. **62**, pp. 106-110, 1998
45. Schott Glass Catalogue, 2003.
46. N. Sugimoto, H. Kanbara, S. Fujiwara, K. Tanaka, Y. Shimizugawa, K. Hirao, "Third-order optical nonlinearities and their ultrafast response in Bi_2O_3-B_2O_3-SiO_2 glasses, J. Opt. Soc. Am. B **16**, pp. 1904-1908, 1999
47. K. Kikuchi, K. Taira, N. Sugimoto, "Highly nonlinear bismuth oxide-based glass fibers for all-optical signal processing", Electron. Lett. **38**, pp. 166-167, 2002
48. N. Sugimoto, Y. Kuroiwa, K. Ochiai, S. Ohara, Y. Furusawa, S. Ito, S. Tanabe, T. Hanada, "Novel short-length EDF for C+L band amplification", Proceedings of Optical Amplifiers and their Applications, Quebec City, Canada, 9-12 Jul 2000
49. Y. Kuroiwa, N. Sugimoto, K. Ochiai, S. Ohara, Y. Furusawa, S. Ito, S. Tanabe, T. Hanada, "Fusion spliceable and high efficient Bi_2O_3-based EDF for short length and broadband application pumped at 1480 nm", OFC 2001, Anaheim, California, 17-22 Mar 2001, TuI5
50. J.S. Wang, E.M. Vogel, E. Snitzer, `Tellurite glass: a new candidate for fiber devices', Opt. Mat. 3, pp. 187-203, 1994
51. R. Stegeman, L. Jankovic, H. Kim, C. Rivero, G. Tegeman, K. Richardson, P. Delfyett, Y. Guo, A. Schulte, T. Cardinal, `Tellurite glasses with peak absolute Raman gain coefficients up to 30 times that of fused silica', Opt. Lett. **28**, pp. 1126-1128, 2003
52. K. F. J. Heinrich, Electron Beam X-ray Microanalysis, (Van Nostrand Reinhold Co., 1981).
53. B. T. Kuhlmey, T. P. White, R. C. McPhedran, D. Maystre, G. Renversez, C. M. de Sterke, L. C. Botten, "Multipole method for microstructured optical fibers. II. Implementation and results," J. Opt. Soc. Am. B **19**, pp. 2331-2340, 2002
54. X. Feng, S. Tanabe, T. Hanada, "Hydroxyl groups in erbium-doped germanotellurite glasses", J. Non-Cryst. Solids **281**, pp. 48-54, 2001

55. A. Boskovic, S. V. Chernikov, J. R. Taylor, L. Gruner-Nielsen, and O. A. Levring, "Direct continuous-wave measurement of n_2 in various types of telecommunication fiber at 1.55μm," Opt. Lett. **21**, pp. 1966-1968, 1996
56. T. M. Monro, D. J. Richardson, N. G. R. Broderick, "Efficient modeling of holey fibers," Proc. Opt. Fiber Commun. Conf. No. FG3, San Diego, California 21-26 Feb 1999
57. W. H. Reeves, J. C. Knight, P.St.J. Russell, "Demonstration of ultraflattened dispersion in photonic crystal fibers", Opt. Express **10**, pp. 609-613, 2000
58. Pottage JM, Bird DM, Hedley TD, et al. "Robust photonic band gaps for hollow core guidance in PCF made from high index glass", Optics Express **11**, pp. 2854-2861, 2003
59. L.B. Shaw, J.S. Sanghera, I.D. Aggarwal, F.H. Hung, "As-S and As-Se based photonic band gap fiber for IR laser transmission", Optics Express **11**, pp. 3455-3460, 2003

ENHANCEMENT OF THE ELECTROLUMINESCENT PHOSPHOR BRIGHTNESS AND STABILITY

M.M. Sychov
St. Petersburg State Institute of Technology
Korablestroiteley 19-1-639
St. Petersburg 199226, Russia
msychov@yahoo.com

ABSTRACT

To improve characteristics of the modern electroluminescent devices it is necessary to use phosphors of smaller grain size, higher brightness and stability as well as better color properties. In the work phosphor particle size was decreased by optimization of the fabrication conditions. Further reduction of the particles size was achieved by the developed fractionation technique based on gravitational sedimentation in the viscous media. Study of the effects of the activator and co-activators on the phosphor's properties allowed us to control phosphor's color properties in a wide range and to improve EL efficiency. Another improvement of the phosphor brightness was achieved by the electron-beam treatment of the charge flux before the sintering step. We assumed that such kind of treatment is likely to provide more even distribution of the activator ions in the phosphor matrix. To examine surface donor-acceptor properties of the phosphors novel technique was used, viz. measurement of distributions of active surface centers (DAC). This method was shown to enable one to obtain data related to the chemical composition of surface, to detect structural defects and basing on that information to predict phosphor's performance.

INTRODUCTION

Electroluminescent powder AC phosphor is a component of electroluminescent panels (ELP) which are utilized for LCD backlighting, as large area light sources for signage, design and advertising etc [1-3].

ELP devices have the following advantages:

- Light weight,
- Flexibility,
- Flat and slim design,
- Absence of heat generation,
- Absence of mercury,
- Good efficiency,
- Low power consumption.

Design of the ELP device is depicted in the fig. 1. Flexible polymer substrate (1) is coated with transparent front electrode (2). Front as well as rear electrode (5) comprises luminescent

(3) and breakdown protecting (4) layers. Luminescent layer is a composite film consisting of polymer binder with phosphor particles dispersed in it.

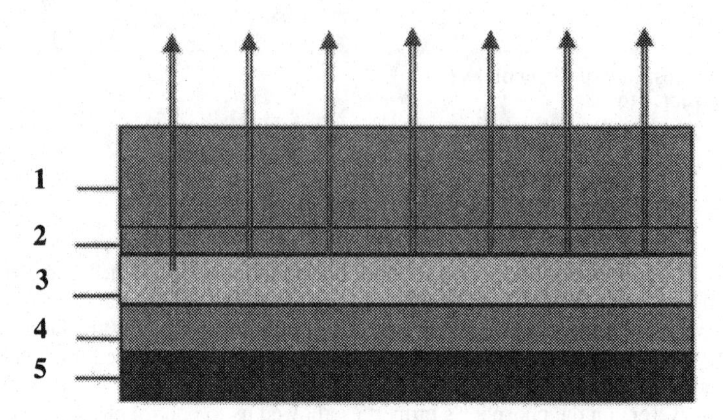

Fig. 1. Schematic structure of ELP device

For the advancement of ELP technology thinner luminescent layers are needed that provides improved efficiency, decreased working voltages and lowered power consumption. To decrease layer thickness and not sacrifice uniformity of the emission smaller particle size phosphors of different colors are needed. In addition phosphors should have good efficiency and stability. This paper is a review of research performed at St. Petersburg State Institute of Technology with the aim to investigate influence of various factors on EL phosphors properties and to control and improve their characteristics. From general point of view study of electroluminescence is more informative method since unlike UV light or electron beam that both have small penetration depth and excite only surface layers of phosphor, electroluminescence involves the whole phosphor particle and thus allows one to judge about bulk properties.

SMALL PARTICLE SIZE EL PHOSPHOR FABRICATION

Firstly we have developed technology of the fabrication of small particle size ZnS:Cu and ZnS:Cu,Al electroluminescent (EL) phosphors. Schematic of the phosphor fabrication process is shown in the fig. 2. To fabricate phosphors high-purity zinc sulfide was mixed with the flux containing copper and (if needed) aluminum compounds and sintered. Sintering was performed in closed quarts crucibles in slightly reducing atmosphere. Charge flux contained sulfur in order to prevent zinc sulfide oxidation and formation of sulfur vacancies. Halogen-containing substance was also added to the flux to facilitate gas-phase process of grain growth. After sintering and consequent annealing, excess of copper sulfide phase was washed out from the surface of the powders. Details of the fabrication process were described elsewhere [4].

To ensure reduction of EL phosphor particle size, initial zinc sulfide was ball-milled, milled powder had average particle size in the range of 1–2μm. Particle size was measured by the computer processing of the SEM images acquired with JEOL JSM-35CF electron microscope. In addition to particle size reduction, milling provided activation of the zinc sulfide surface that is favorable for the promotion of interfacial processes during phosphor particle growth [5]. Sintering temperature was lowered to less than 1000°C to limit particle growth. Grains of fabricated phosphors were of quite regular shape and particle size distribution was relatively narrow. Average particle size was 14μm comparing to 25μm for the commercial sample – fig. 3.

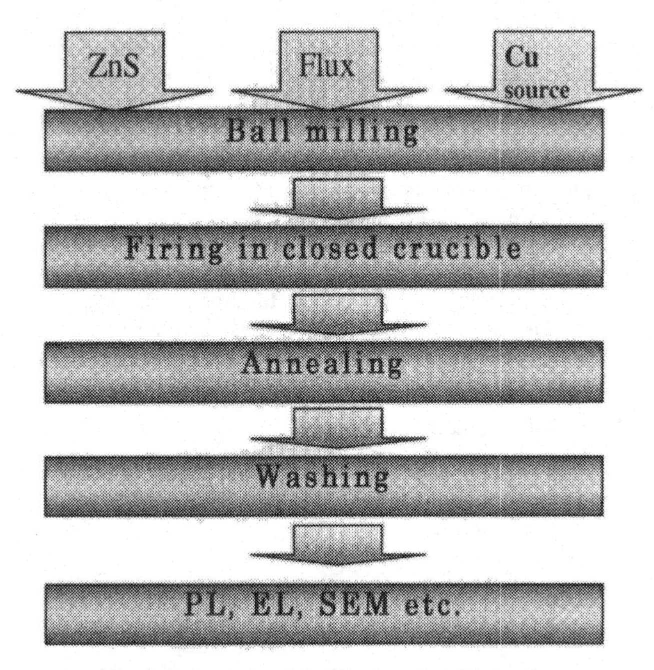

Fig. 2. Schematic of the EL phosphor fabrication

Fig. 3. SEM image of the EL phosphor particles

For the purpose to further decrease phosphor particles size we have developed fractionation technique based on gravitational sedimentation in the viscous media. Conditions were found which allowed us to separate small particles size fraction of the phosphor (0 to 10μm). Initial and fractionated samples were compared in brightness and stability (defined as half-brightness time in the accelerated aging tests). To measure EL characteristics electroluminescence panels having structure shown in fig.1 were prepared using butadiene acrylic-nytrille rubber as a binder. Brightness was measured with IL 1700 radiometer under the 100V and 400Hz, data from 4 to 6 samples was averaged to ensure reliability. The brightness of EL phosphor with small particle size was twice as much as that of initial sample while its stability was almost the same [6].

ACTIVATOR CONTENT EFFECT ON THE PHOSPHOR PROPERTIES

Effect of the copper content on phosphor's properties was investigated on the example of ZnS:Cu,Br phosphor. Copper participates in two important processes related to electrolumnescence. First of all it forms copper sulfide precipitates in zinc sulfide matrix. When external electrical field is applied to phosphor particle, resulted $ZnS-Cu_2S$ heterojunctions generate charge carriers needed to produce luminescence. Copper also participates in the formation of luminescence centers themselves where charge carriers recombination and light generation takes place.

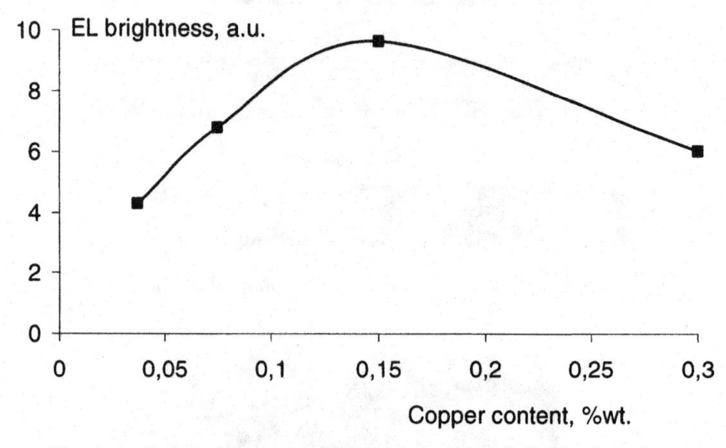

Fig. 4. Activator effect on the EL intensity of ZnS:Cu,Br phosphors

We have studied copper effect by varying its content in the charge mixture from 0.0375 to 0.3000 wt.%. It is well known that copper increase lead to the brightness rise for this type of phosphors [3] but we observed somewhat different behavior. As one can see in fig. 4, EL brightness reaches its maximum at a certain copper content. The increase in brightness is caused by formation of additional luminescence centers along with the shift of emission toward the yellow region of visual spectrum to which human eye is more sensitive – see fig. 5. However when copper content in the phosphor exceeds certain optimum value, brightness goes down. Since this decrease can't be explained by the change of spectra it should be attributed to the decrease of phosphor's efficiency. Two main factors are responsible for that. Firstly increased copper content yields higher amount of light-absorbing Cu_2S phase. And secondly concentration quenching takes place at high copper contents [1]. Thus for the best phosphor efficiency it is important to optimize activator content.

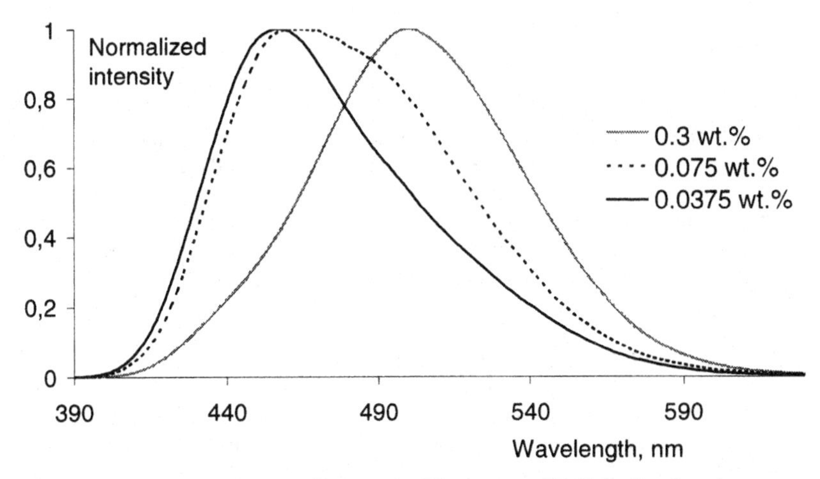

Fig. 5. Copper content effect on the EL spectra of ZnS:Cu,Br phosphors

ELECTRON BEAM TREATMENT OF THE STARTING ZINC SULFIDE

From the results represented above it is clear that while increased copper content has positive effect on the phosphor brightness due to formation of additional luminescence centers, there is also negative effect due to precipitation of the excess copper sulfide phase and related decrease of the phosphor's efficiency. It was assumed then if we increase copper content in the phosphor and simultaneously improve uniformity of its distribution within the ZnS matrix (in other words its solubility) it would provide positive effect without much of the negative one and would in turn allow us to increase EL brightness above the achieved maximum. To test this approach we subjected starting zinc sulfide to electron-beam (EB) treatment before the sintering. The idea was that it would result in formation of dislocations that in turn would make diffusion of the copper into ZnS matrix easier and provide its more uniform distribution within growing particle during phosphor synthesis. For electron-beam treatment of phosphor powders a transformer type electron accelerator RTE-1V was utilized. The treatment was conducted under the following conditions: 1mA current, 900 keV electron energy, 500 kGy absorption doze.

The approach was checked up for the ZnS:Cu,Br phosphor. Fabrication of both phosphors – from untreated (sample 1) and EB treated ZnS (sample 2) was performed simultaneously in the same oven to exclude influence of technological factors. In the both cases amount of copper in the charge mixtures was the same.

Table I. Effect of EB treatment of ZnS on phosphor's characteristics

Characteristic	Value		Change,
	Sample 1	Sample 2	%
Copper content, ppm	750	890	+20
Brightness at 100V drive, a.u.	14	17	+20
Brightness at 220V drive, a.u.	56	71	+25

As expected sample 2 had increased amount of activator (~20% rise as measured by atomic adsorption spectroscopy) – table I. That led to a 20-25% increase in the EL brightness. For the higher drive voltage increase of brightness is greater presumably because at higher voltages smaller particles which have higher efficiency mostly input into the overall emission while at low voltages the large ones. Having larger specific surface and thus area of the contact with copper containing flux small particles contain more copper due to its increased diffusion.

CO-ACTIVATOR EFFECT ON THE PHOSPHOR PROPERTIES

Co-activator effect is also very important in the optimization of the phosphor properties as it affects particle growth, copper content etc. Phosphor samples with different halogen co-activators namely chlorine, bromine and iodine were fabricated. Their EL spectra are presented in the fig. 6.

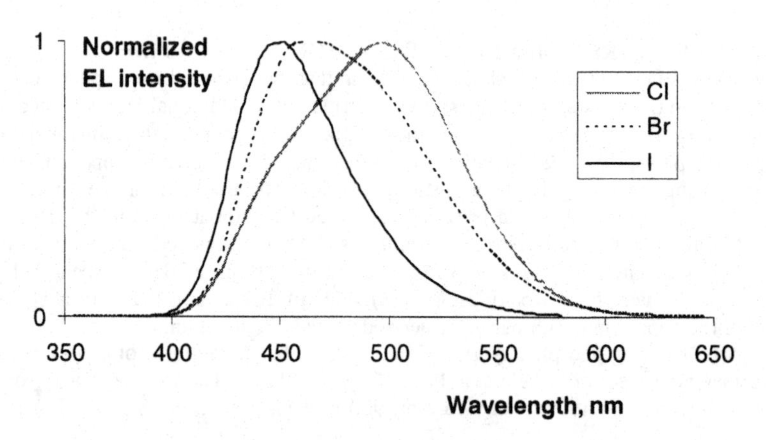

Fig. 6. Co-activator effect on phosphor's spectrum

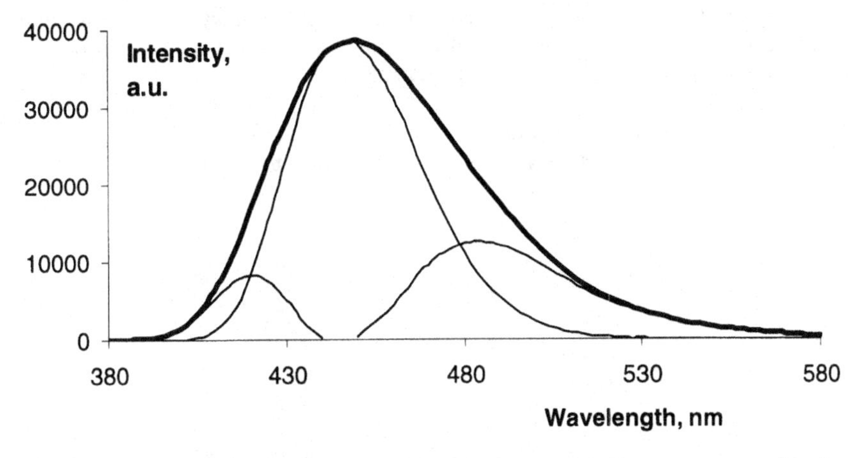

Fig. 7. Example of decomposition of ZnS:Cu,Br phosphor luminescence spectra into its components: bold line – luminescence spectra, thin lines – bands

A shift of spectra is observed from green to blue region when halogen was changed from chlorine to iodine. That is due to the change of the relative intensities of bands in emission spectra. We have decomposed luminescence spectra of samples using Fok-Alentsev method [7]. Three bands – "green", "blue" and "violet" – were identified on the basis of attribution made in the work [8]. Attribution of bands exemplified in the fig. 7 is presented in the table II. "Green" band is formed by luminescence of donor-acceptor pairs comprising associates of the copper ion substituting zinc site in the lattice and halogen ion substituting sulfur site. "Blue" band is related to the donor-acceptor pairs comprising associates of the copper on the zinc site and interstitial copper. "Violet" band appears owing to the energy transitions between the conduction band and energy levels of vacancies.

Table II. Attribution of emission bands

Band	Peak wavelength, nm	Origin
Green	505	$[Hal_S]^d$-$[Cu_{Zn}]^a$
Blue	455	$[Cu_i]^d$-$[Cu_{Zn}]^a$
Violet	405	Sulfur vacancies

Thus the cause of spectra change is the decrease of the solubility of the co-activator in the zinc sulfide with the ion radius increase. As a result solubility of the copper in the phosphor also diminishes as halogen provides charge compensation for the incorporated copper. That in turn leads to the decrease of relative amount of "green" luminescence centers and ratio of the intensities of "green" and "blue" bands in the EL spectra falls. Fig. 8 illustrates the situation – if for the phosphor co-activated with chlorine this ratio is more that 5 (absolutely prevailing are "green" centers), for the case of iodine the ratio falls below 1 that means "blue" centers become dominating. In the same time intensity of the "violet" band attributed to the presence of sulfur vacancies increases. That is also understandable since increased ion radius of halogen prevents it from efficient occupation of sulfur vacancies.

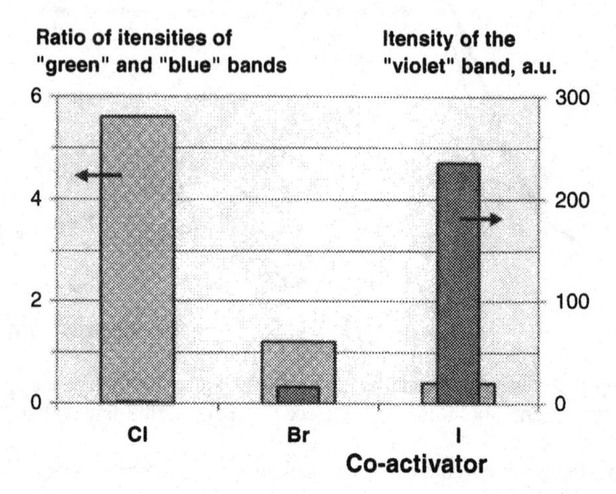

Fig. 8. Effect of the co-activator on spectral properties of phosphors

However nature of halogen affects not only chemical composition but also particle size of the phosphor as found form SEM data shown in table III. Computer processing of the images with the use of ImageTool 2.0 software allowed us to measure ranges and average values of minimum and maximum sizes of phosphor particles. Due to the easier decomposition of the halogen-containing compound in the Cl<Br<I set in sintering atmosphere, gas-phase reactions and mass transfer intensifies and particle growth is accelerated.

Table III. Co-activator effect on phosphor particle size

Co-activator	Particle size, µm			
	d_{min}	d_{min} range	D_{max},	D_{max} range
Cl	2.4	0.2 – 7.5	3.4	0.3 – 15
Br	3.2	0.4 – 18	4.6	0.4 – 21
I	6.0	0.6 – 28	8.5	0.9 – 35

The results obtained in the work showed that EL brightness had the highest value in case of phosphor co-activated with bromine (see fig.9) due to the superposition of two discussed above effects – decreased copper content and increased particle size when chlorine-containing compound in the charge mixture is exchanged to bromine and to iodine.

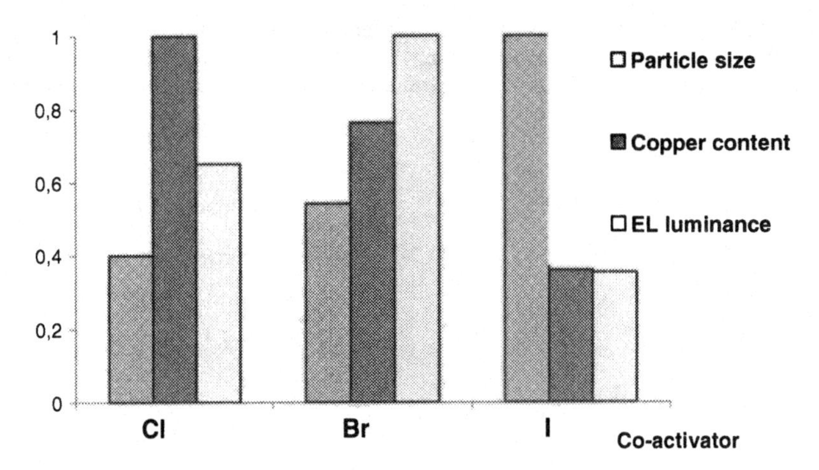

Fig. 9. Normalized properties of phosphors with different co-activators

SURFACE PROPERTIES AND PERFORMANCE OF PHOSPHORS

Since surface of the phosphor play important role in the luminescence processes it is crucial to investigate and be able to regulate its properties. In this paper we studied surface properties of phosphor powders by the distribution of active centers technique (DAC). DAC method is based on the measurements of amounts of specific surface centers (qpK_a value, μmol/g) with certain values of pK_a – dissociation constant characterizing acid-base nature of the center [9]. Moreover both differential (distributions) and integral (total amount of centers, acidity function H_o) characteristics of surface may be obtained.

Previously it was shown that for the ZnS:Cu phosphor with different copper content, amount of specific centers namely weak acid ones around pK_a=3-4 have very strong positive correlation with amount of the copper in one order of magnitude range of copper concentrations [10,11]. In other words it was shown that DAC allows one to study chemical composition of the surface very preciously. Further we studied peculiarities of the donor-acceptor spectra of surface centers and tried to find correlations with the technology and performance of the phosphors.

First of all effect of sulfur in the sintering atmosphere was studied. As one can see from the fig. 10, introduction of sulfur into the flux led to the drastic decrease of the basic centers with pK_a in the range of 11–13. These centers were attributed to the Zn–OH surface groups formed next to sulfur vacancies. This attribution is supported by the results of spectroscopic studies: intensity of the "violet" band in the EL spectra which was connected to the presence of sulfur vacancies decreased more than one order of magnitude when sulfur was introduced into the charge flux. Amount of the centers with pK_a=2.5 (weak Broensted acid centers) also decreased significantly. Those centers were attributed to the presence ZnO phase according to results presented in the work [12]. It is quite reasonable that introduction of the sulfur having reducing properties prevents formation of zinc oxide during the sintering.

Annealing of the phosphor provided the same effect of decrease of amount of strong Broensted basic centers and increase of the EL brightness – fig. 11. One can see that brightness increased with longer annealing, while amount of defect-related centers decreased since heat-treatment promotes diffusion and improvement of the crystal structure. Finally, described above EB treatment of the initial zinc sulfide before the sintering also gave the same positive effect.

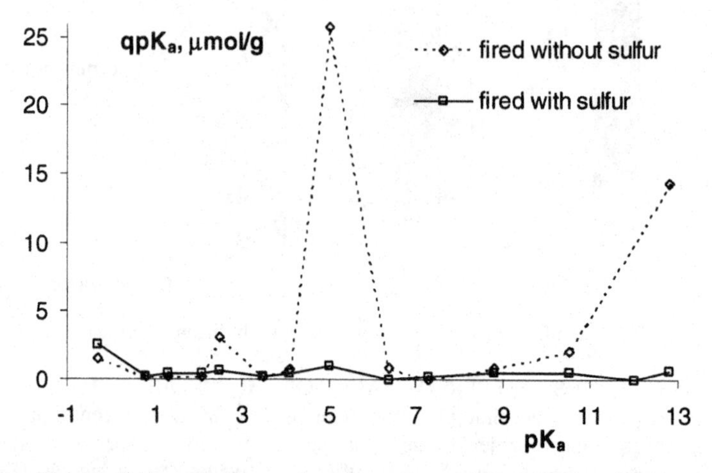

Fig. 10. Sulfur effect on surface properties of ZnS:Cu,Al phosphors

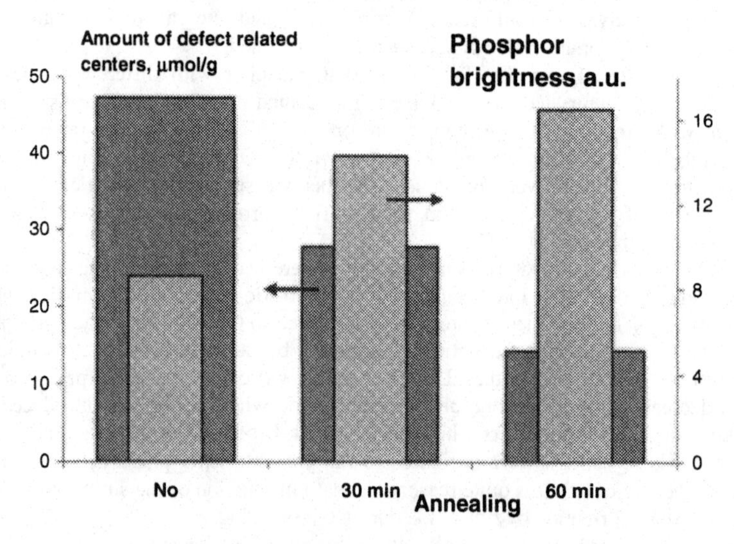

Fig. 11. Annealing effect on phosphor's properties

Thus DAC method gives the possibility of precious study of the chemical composition of phosphors surface as well as estimation of its quality and detection of surface structural defects. Of cause method may be applied correctly only to the series of samples having similar origin and technology of fabrication. Nevertheless it proved to be very useful in the phosphor technology as it provides control of phosphor's quality and even prediction of its performance.

CONCLUSIONS

- The developed technology of the electroluminescent phosphor fabrication yields small particle high brightness and stability material. Study of the activator and co-activators effect on the phosphor's properties allowed to maximize efficiency and control color properties in a wide range. Figure 12 reflects color gamut of the developed phosphors.
- Electron beam treatment of the initial zinc sulfide powder provided substantial increase of phosphor brightness. Observed fact was referred to increased copper solubility and uniformity of its distribution within zinc sulfide matrix.
- It was shown that presence of sulfur in the sintering atmosphere as well as annealing of fabricated phosphors decreased amount of structural defects and gave rise to increased EL brightness.
- New method for the measurement of the chemical composition of the surface of powder phosphor allowed to estimate quality of the surface in terms of amount of sulfur vacancies. While applicable correctly only to the series of samples having similar origin and technology of fabrication, it proved to be useful by provision of means to control phosphor's quality and even predict its performance. In addition DAC technique enables characterization of surface chemical composition. Thus it may be useful as QC/QA component in the phosphor technology.

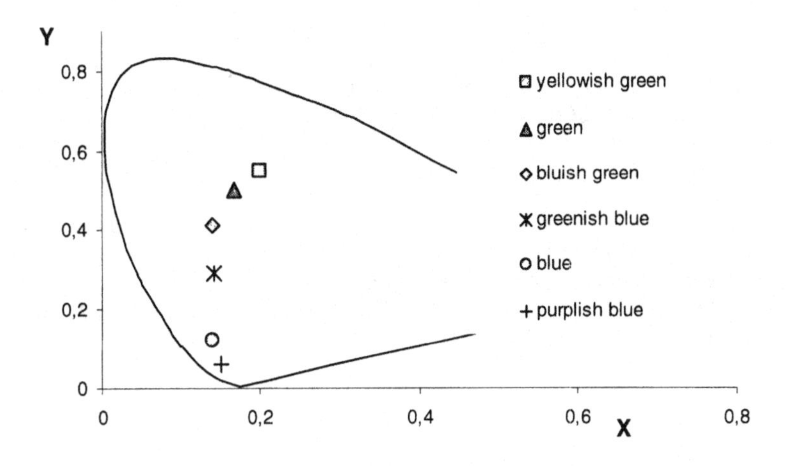

Fig. 12. Color coordinates of the developed EL phosphors

ACKNOWLEDGEMENTS

Author is very grateful to all colleagues and students contributed into presented research, especially V.G. Korsakov, V.V. Bakhmetiev, E.V.Komarov, S.V.Mjakin and I.V.Vasilieva. Financial support of St.Petersburg Committee on Science and Higher Education (concurs of year 2004 on applied chemistry projects) is greatly appreciated.

REFERENCES

[1] Lai Qi, Burtrand I. Leea, Xuejun Gu, Mica Grujicic, W.D. Samuels, G.J. Exarhos, "Concentration Efficiency of Doping in Phosphors: Investigation of the Copper- and aluminum-doped zinc sulfide," *App.Phys.Let.* 83 [24] 4945-4947 (2003).

[2] D.A. Davies, J. Silver, A. Vecht, "Particle Size Effects in AC Powder Electroluminescence," *Proc. of I Int. Conf. on Sc. and Tech. of Disp. Phosphors,* 163 (2001)

[3] M.V. Fok, "Applied Electroluminescence," Radio, Moscow, p.62 1974.

[4] M.M. Sychov, V.V. Bakhmet'ev, S.V. Mjakin, "Modification of the EL Properties of Zinc Sulfide Powder Phosphors," *Proc. of Electronic Display Conference,* Nagasaki, p.67 (2002).

[5] V.V. Bakhmet'ev, E.N. Kalininina, M.M. Sychev, V.G. Korsakov, O.A. Cheremisina, "Surface Properties of ZnS and phosphors based on it," pp.61…65 in *New Investigations in the Materials Science and Ecology,* PGUPS, St. Petersburg, 2001.

[6] M.M. Sychov, V.V. Bahmet'ev, L.V. Khavanova, A.I. Kuznetsov, A. Smirnov, I.V. Vasil'eva, S.V. Mjakin, Y. Nakanishi, "New Methods in the Technology of Electroluminescent Phosphors," *Proceedings of IMID-2003,* Daegu, Korea. 1065-1070 (2003).

[7] M.V. Fok, "Decompositon of Complicated Spectra onto Individual Bands," *Proc. of Phys. Inst.* **59** 3-23 (1972).

[8] A.G. Miloslavsky, N.B. Suntsov, "Defect Structure and Luminescenec Centerss of Zinc Sulfide Phosphors," *Phys. And Tech of. Semic.* **7** [2] 94-103 (1997)

[9] O.A. Cheremisina, M.M. Sychev, S.V. Myakin, V.G. Korsakov, V.V. Popov, N.Yu. Artsutanov, "Dispersing Effect on the Donor-Acceptor Properties of the Surface of Ferroelectrics," *Rus. J. of Phys. Chem.* **9** [76] 1472-1475 (2002).

[10] M.M. Sychov, V.V. Bakhmet'ev, Y. Nakanishi, S.V. Mjakin, L.V. Havanova, O.A. Cheremisina, V.G. Korsakov, "Surface properties of ZnS and AC powder electroluminescent phosphors," *Journal of the SID* **11** [1] 33–38 (2003)

[11] M.M. Sychov, Y. Nakanishi, V.V. Bakhmet'ev, V.G. Korsakov, N.M. Sergeeva, N.V. Zakharova, S.V. Mjakin, I.V. Vasil`eva, "Control of EL powder phosphor properties", *SID International Symposium,* Boston 400-403 (2002).

[12] A.P. Nechiporenko, "Donor-Acceptor Properties of Surface of Solid Oxides and Halcogenides," Thesis of the Dr.Sc. dissertation. St.Petersburg, 1995.

FEMTOSECOND LASER INDUCED STRUCTURAL MODIFICATION AND BIREFRIENGENCE IN BULK GLASS FOR OPTICAL WAVEGUIDE APPLICATIONS

Pin Yang, David R. Tallant, George R. Burns, and Junpeng Guo
Sandia National Laboratories
Albuquerque, NM 87185-0959

ABSTRACT

Embedded waveguides and their optical properties in amorphous silica fabricated by femtosecond (fs) laser pulses (800 nm, < 125 fs, at 1 kHz) are reported. Experimental results show that there is a narrow operating window to produce low loss waveguides. Raman spectroscopy of the laser-modified glass regions shows a reduction of large network silica rings in the glass, which might contribute to the local densification and refractive index change. Several integrated optical devices such as a Y coupler, directional coupler, and Mach-Zehnder interferometer made by this technique will be presented. An angular dependence of light transmission measured between two cross polarizers on these laser-modified regions suggests that these regions possess an optical birefringent property. Furthermore, the optical axes of laser-induced birefringence can be controlled by the polarization direction of the fs laser. Mechanisms that contribute to the observed laser induced birefringence behavior are discussed.

INTRODUCTION

The development of an ultrahigh peak power laser pulse up to the terawatt (TW) level by a chirped pulse amplification technique[1] opens up new possibilities for micro-fabrication of three-dimensional (3D) features inside of transparent materials. These compressed laser pulses (with pulse widths less than 1 pico-second), typically are not absorbed by the work media in the near-infrared range (800 nm) through a single photon process. However, the energy of these pulses can be absorbed at the focal point through a non-linear, multi-photon absorption process when local power density (or irradiance) exceeds TW per cm^2.[2] This localized absorption changes the physical and chemical properties of the transparent media and permits direct-write 3D micro-structuring by scanning the focal spot inside of the materials. Using the femtosecond (fs) laser induced refractive index changes in glass, embedded waveguides[3,4] and other optical devices[5,6,7] fabricated in three dimensions have been demonstrated. In addition, recent literature demonstrated that the fs laser irradiated areas exhibit high etching selectivity, which allows the fabrication of embedded channels[8,9] in glass. The combination of these new capabilities created by the fs laser-induced material modification provides new opportunities to fabricate novel integrated optics and sensor systems.

Embedded waveguides can be created by moving the focal spot with proper pulse energy inside of a bulk glass. These waveguides can be directly written either transversely or longitudinally to the beam direction. Waveguides written by a longitudinal approach[1,10] have a symmetrical cross-section. However, direct writing a 3D waveguide inside of a glass is limited by the focal length of the microscope objective. The transverse writing approach is convenient and has less constraint imposed by the focal length of the microscope objectives, but the intensity profile at the focal spot is elongated, which creates a highly asymmetric cross-section at the laser-modified region.[11] Approach to reduce the asymmetric beam profile, either by introducing beam shaping optics[11] or a parallel slit[12], has shown great success. In this paper, we used a transverse writing approach to create waveguides in amorphous silica, without introducing additional beam shaping processes. The properties of these waveguides with respect to the laser processing conditions are reported. A few prototype devices were fabricated to demonstrate the waveguide quality and the flexibility of this direct-write technology.

Two different approaches have been developed to deposit the photon energy into glass and induce localized refractive index change. The first approach uses a high-repetition-rate (MHz range), low-peak-power laser pulses which have just enough laser intensity (power per unit area) to create multiphoton absorption in glass. This approach generates localized bulk heating[13] and densification, which creates an isotropic refractive index change in bulk glass. The second approach employs a low-repetition-rate (KHz range), high-peak-power laser pulses. This approach breaks chemical bonds, generates nonbridging oxygen hole centers,[4] modifies the silica network,[14] and creates localized refractive index change in glass. It is also found that this high-peak-power pulse approach can induce unusual birefringent properties[6,15,16] in laser-modified regions. The fundamental mechanisms for this unusual birefringence created by the fs laser pulses in the bulk glass are not clear. In this paper, experimental evidence that could contribute to the laser-induced birefringence in bulk glass is presented and discussed.

EXPERIMENTAL PROCEDURE

The laser used in this experiment is a regeneratively amplified Ti:Sapphire femtosecond pulsed laser (Spectra-Physics, Hurricane). The fs laser pulses have a nominal 125-fs pulse-width (FWHM) and 1 kHz pulse repetition rate. The energy of the pulse can be varied by the combination of a half wave plate and a thin film polarizer. Further energy reduction is accomplished by the use of neutral density filters to avoid photo-damage to the polished glass plates. The average laser power used in this investigation, without creating photo-damage to the glass, is between 450 μW and 900 μW which corresponds to a pulse energy between 0.45 μJ and 0.9 μJ. The laser is focused to a 3 μm diameter spot inside the glass, using a long-working-distance microscope objective (Mitutoyo, 20X, NA = 0.40). The calculated peak intensity of the fs pulse is $9.8 \text{ X}10^{20}$ W/m^2 for a pulse energy of 0.9 μJ.

In this investigation, a thermally treated chemical vapor deposited (CVD) amorphous silica plate (Quartz International, Albuquerque, NM) was used. These glass plates are optically isotropic before laser writing. The glass plate (polished on both surfaces) was mounted on a motorized x-y translation stage with linearly controlled motion along the z-direction to adjust the focus of the laser inside of glass. The laser was focused approximately 600 μm below the surface of the glass plates. In addition to these linear motion controls, the sample can also be rotated in the x-y plane with a rotational stage. For the birefringence study, the polarization direction of the beam was fixed and the sample was rotated with respect to the polarization direction of the laser beam. By moving the laser focus spot perpendicular to the beam direction (20 μm/s) with an appropriate energy level (0.45 to 0.9 μJ per pulse) inside of bulk glass, optical waveguides can be created. The general quality of these waveguides was studied by near-field and far-field methods.

The laser-induced local structural modification was analyzed, using the microscope accessory to the Raman spectrometer and the 458 nm laser line as an excitation source. A microscope objective focused the excitation laser into the glass and collected 180° backscattered light for dispersion and detection by the spectrometer. Collected light was dispersed by a triple mononchromator and detected by a charge-coupled detector. The volume sampled using this technique approximates a cylinder of 1 μm diameter and 3-6 μm in depth, which is perpendicular to the propagation direction of the laser beam.

After laser writing, the samples were evaluated using a transmission microscope (Olympus BH-2). White light was used from a halogen lamp. The samples were put between two crossed polarizers on the rotation stage of the microscope. The angular rotation of the sample with respect to the polarizer was accurately measured by the stage goniometer. The intensities of the transmitted light through the samples were measured by using a CCD digital camera attached to the transmission microscope viewer. Optical transmission signals were captured using a National Instruments PCI-1424 digital image capture card. The CCD camera is operated in the linear regime. The maximum intensity of the transmitted light for the laser-modified region was determined by scanning the image array for the maximum value in each of ten images. The value was then averaged and recorded.

RESULTS

Processing Condition and Waveguide Properties

Due to the nonlinear absorption processes, it is important to identify the proper processing space that can effectively change the refractive index without damaging the glass. It was found that the threshold of pulse energy that can create a detectable waveguide pattern by the nonlinear absorption processes is above 0.2 μJ. Embedded waveguides without significant scattering loss can only be fabricated with pulse energy between 0.45 to 0.90 μJ, [17] which is consistent with experimental observations reported in the literature.[14] Above this energy level, stress birefringence and sometimes microcracks can be observed, indicating photo-damage to the bulk glass. Fig. 1 shows laser-induced microcracks and stress-induced birefringence on the edges and at corners of a square test pattern under cross-polarized light, when pulse energy is much greater than 1 μJ/pulse. In contrast, two spots (near the center) created by an accumulated 60,000 pulses at 0.45 μJ/pulse exhibit no detectable cracks or stress birefringence. Experimental results indicate that the photo-damage threshold in silica is extremely sensitive to the change of laser pulse width.

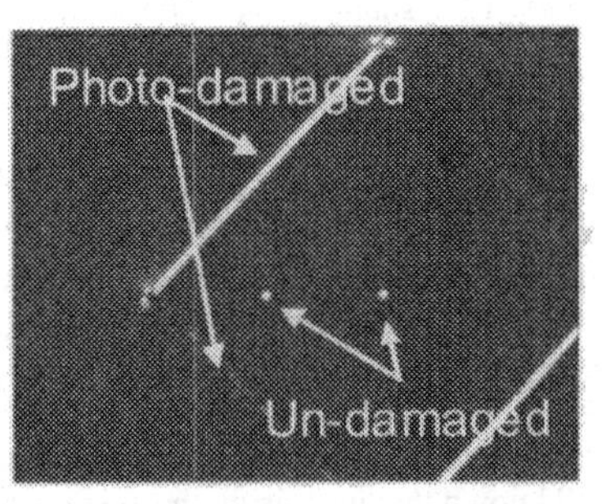

Fig. 1 Microphotograph of laser damaged and undamaged regions in silica. (2.5 mm X 2.5 mm square)

Difficulties were experienced when creating refractive index changes on the glass surface. Attempts using different optics and power levels to write surface waveguides were unsuccessful. It was found that a minimum optical penetration depth is required. Results show that embedded waveguides or refractive index modification can be created when focal point is below 15 μm from the surface of an amorphous silica plate. Because the band gap of silica is large (~ 9.0 eV), the probability for multiphoton absorption directly from 800 nm wavelength photons (E = 1.55 eV) is minute since it requires a stacking of 6 photons to cross the band gap. It is believed that laser energy is absorbed near the focal spot by two to three high energy photons created by the blue-shift of a self phase modulation when intense laser pulses propagating through an optical media.

Fig. 2 illustrates the changes of transmission mode for laser direct-write waveguides in amorphous silica as a function of processing conditions. Light from a laser source (λ = 650 nm) was coupled into the waveguide through a focusing lens, and the near-field intensity distribution of the guided mode was obtained by imaging the backside of the fabricated waveguide onto a CCD camera. By carefully aligning the axis of the input laser and the orientation of the waveguide (~ 3 μm in diameter), a single mode (TEM_{00}) can be excited in the waveguides fabricated with a pulse energy of 0.45 μJ. This single guided mode with a near Gaussian profile is demonstrated by a 3D field intensity distribution on the insert of Fig. 2. However, when the number of passes increases (as shown in the X axis) or the pulse energy increases (in the Y axis), a higher-order mode (TEM_{01}) is developed in these waveguides. Finally, when the pulse energy exceeds the photo-damage threshold, high-order, multiple-mode (TEM_{11}) is observed. Furthermore, most of the light was scattered through propagation, and the damaged waveguides did not efficiently guide light.[17] It is well known that increasing the average laser power can enlarge the laser-modified region[18] and increasing the number of accumulated laser pulses can enhance the refractive index change[19] in the glass. The combination of creating waveguides with a larger diameter and a higher refractive index change alters the transmission property from a single guided mode to high-order, multiple modes. Similar results have also been reported by changing the writing speed, where decreasing writing speed switches the guided mode from a single fundamental mode to higher order modes.[20]

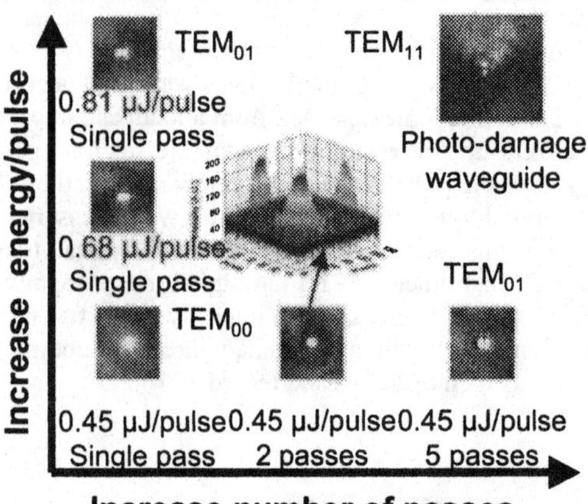

Fig. 2 Processing condition versus transmission mode of laser direct-written waveguides, measured by near-field images.

The control of transmission loss in the waveguides is important for practical applications, which is extremely sensitive to the fabrication process, glass composition, and defects in the optical medium. For a straight waveguide, the majority of the loss can be attributed to the light absorption by the material and the light scattering from defects along the light propagation direction. The overall transmission loss can be determined by the amount of light scattering along the propagation direction, by measuring the light intensity variation perpendicular to the writing direction. High quality waveguides will show small or undetectable intensity variations along

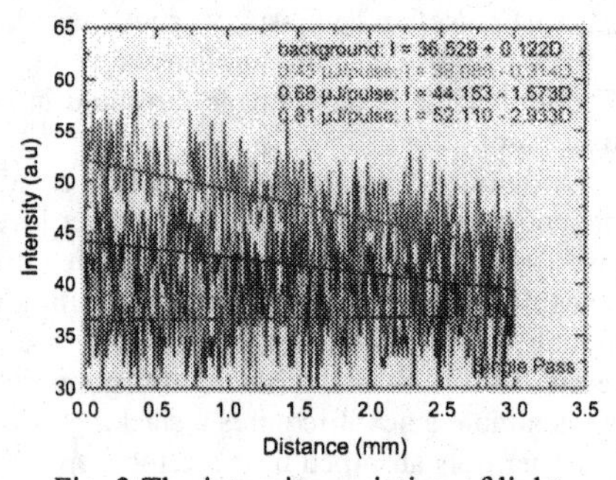

Fig. 3 The intensity variation of light scattering in the waveguide along the propagation direction.

the light transmission direction. Fig. 3 illustrates the light intensity changes (due to scattering) along the transmission direction in laser direct-written waveguides as a function of pulse energy, after light was first coupled into these waveguides and followed by self mode adjusting. These straight lines are obtained from a linear fitting of light intensity with respect to the propagation distance. Results show that waveguides fabricated with higher pulse energy have stronger light scattering and quick intensity decay along the transmission direction (as indicated by the negative slope change). The losses in these waveguides, based on the light scattering measurement (with light source at 650 nm), vary from 0.36 dB/cm to 3.59 dB/cm as pulse energy increases from 0.45 to 0.81 μJ/pulse. However, the transmission loss determined by the near-field intensity for waveguides (fabricated at 0.45 μJ/pulse) of different lengths is about 1.13 dB/cm, which is similar to data in the literature.[20]

The alteration of refractive index between the laser modified core region and surrounding glass can be roughly estimated from the numerical aperture of the fabricated waveguide by a far-field measurement. The estimated value for the refractive index change measured at the onset of the double-lobed far-field pattern with pulse energy at 0.45 μJ/pulse is about $1.34 - 1.57 \times 10^{-4}$. Although a higher refractive index change can be produced with increased pulse energy, the transmission properties of the waveguide will be forfeited. As a result, the following optical devices were made at the pulse energy of 0.45 μJ.

Prototype Optical Devices

The use of integrated optics for sensing applications is receiving increasing attention. Optical methods of transduction offer high sensitivity, and integrated optics have the advantage of a compact structure combined with the potential for detection of several analytes, simultaneously. A planar directional coupler has been demonstrated for sensing biochemical species[21] such as pollutants. The new capability to fabricate waveguides inside of amorphous silica by fs laser pulses has fueled a major thrust to explore this technology for device fabrication. This approach simplifies the manufacturing process and provides greater flexibility compared to conventional waveguide fabrication processes. A variety of devices, such as waveguides[3, 4, 14, 17, 19, 20], gratings[22], couplers[23], splitters[5], micro-mirrors[24], and photonic structures[7] have already been realized using this method. Sandia National Laboratories also has a great interest to develop this direct-writing approach to fabricate integrated optics. Several prototype devices fabricated at Sandia are presented.

Y coupler:

Fig. 4 shows the intensity variation of the outputs in a photo-written Y coupler, based on the schematic layout in the insert. The written pattern has a 2° bend on each arm and a 155 μm separation. Using a pulse energy of 0.45 μJ and a writing speed of 20 μm/sec, the intensity of outputs from each arm in this Y coupler is almost equally divided (50/50 splitting). The image of the output (see the insert image) shows high intensity spots at the end of each waveguide and a weak interference pattern between these two arms. The output intensity was found to be strongly dependent on the bending angle. Results

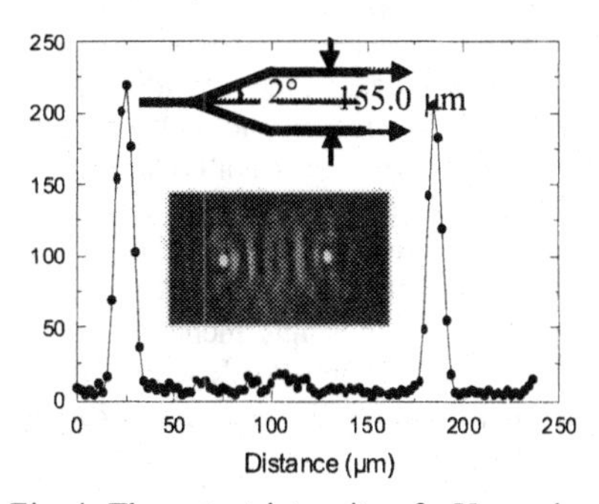

Fig. 4. The output intensity of a Y coupler.

indicated that when the bending angle increased to 10° hardly any light can be transmitted through this bending configuration.

Directional Coupler:

Fig. 5 illustrates the output intensity differences between a long bent waveguide and a coupled waveguide. The schematic of this directional coupler is shown on the right hand side of the figure. The coupling property was studied by focusing 650 nm laser beam into the long waveguide and guided through two 2° bends and observing the near-field pattern of the coupler's output. The output of this launching waveguide is shown as the brightest spot on the image (see inserted image), with an intensity that saturated the light detector (as shown by a flat top on Fig. 5). Along the straight launching direction, a far field pattern was generated. A weak, yet distinct light spot can also be observed to the left of the brightest spot in this image. This spot represents the light intensity created by an electromagnetic coupling through a 3 μm gap between the launching and coupled waveguides. The shape of these spots is indicative of single-mode propagation.

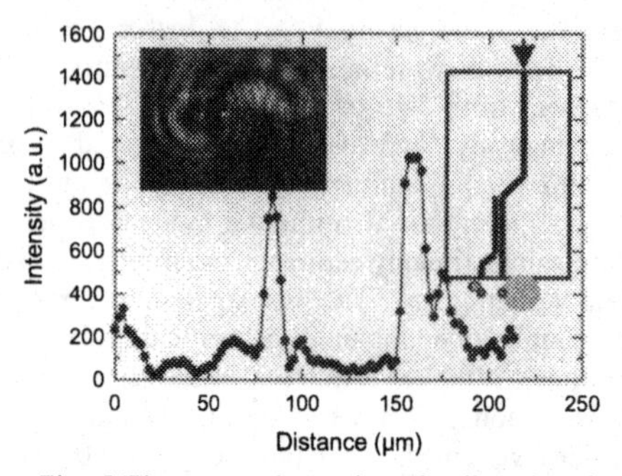

Fig. 5 The output intensity of a directional coupler. The intensity of the launching waveguide is saturated.

Mach-Zehnder Interferometer

Fig. 6 shows the changes of output intensity of an equal path-length Mach-Zehnder interferometer as a function of time (t_o to t_4 as time increases) when one arm of the interferometer was heated by a soldering iron. The schematic of the directly written optical interferometer is inserted on the top of the figure. Local heating generated from the soldering iron changes the refractive index of the waveguide and creates a phase shift between two arms, which alters the intensity of interference pattern. Therefore, as glass temperature near the soldering iron progressively increases, the amount of phase shift increases and the overall output intensity due to intereference from both arms decreases.

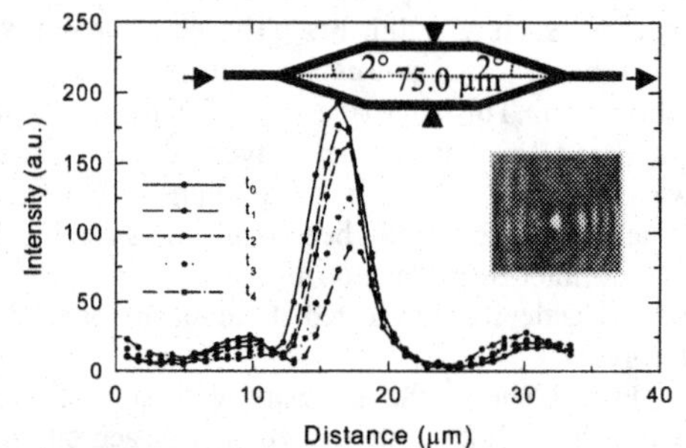

Fig. 6 Progressive change of the interference intensity output as one of the arms in this interferometer is heated.

In this section, we have demonstrated that quality waveguides and optical devices can be created by direct-writing of fs laser pulses. The construction of these devices relies on the ability to create low loss, single mode waveguides in the bulk glass. These results demonstrate that direct-writing of embedded waveguides by fs laser pulses has progressively moved from a laboratory curiosity to a mature fabrication process.

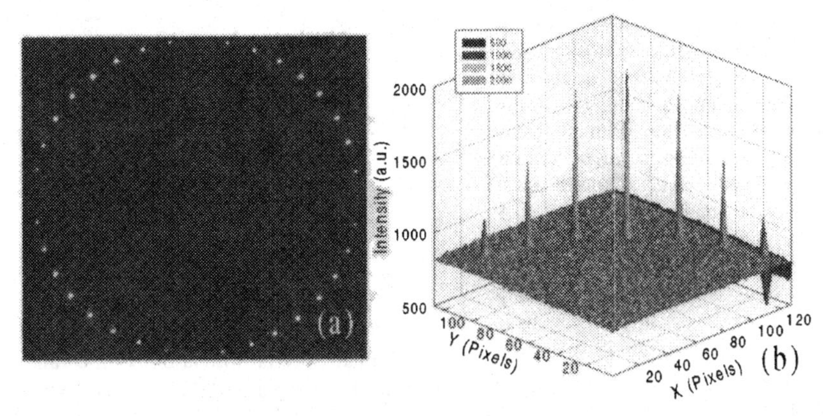

Fig. 7 (a) The transmission image of laser modified region under a cross-polarized light condition, the polarization of the laser beam is rotated 10° for each modification, (b) the change of transmitted light intensity in the first quarter of the circle.

Fig. 7 (a) illustrates the variation in transmitted light intensity through laser-modified regions between two cross-polarizers. Each modified region was exposed to 60,000 pulses (1 minute dwell time) parallel to the beam direction with a pulse energy of 0.9 μJ/pulse. The polarization direction of the laser beam was systematically rotated 10° from spot to spot, by rotating the sample with respect to the polarization direction of the laser beam. The observation of an optical extinction for every 90° suggests that the laser-modified regions possess an optical birefringence property. Furthermore, the sinusoidal variation of the transmitted light intensity (see Fig. 7 (b)) is consistent with the theoretical prediction of a birefringent material.[15] Results indicated that the laser-induced birefringence property in the bulk glass can be controlled by laser power level, accumulated exposure, and polarization direction of the writing laser. Details of the quantitative measurement of birefringence created by the fs laser pulse were reported elsewhere.[15]

DISCUSSION

Laser-Induced Structure Modification in Glass

To understand the effect of structure modification by fs laser pulses in the embedded waveguide regions (below the photo-damage threshold), laser-modified dots with 5,000 accumulated pulses were created with different pulse energy for a Raman spectrometer study. The laser-modified regions are the six patterns of four dots (viewed end-on) in a square, each square corresponding to a different pulse energy (0.90, 0.81, 0.72, 0.63, 0.54, and 0.45 μJ/pulse), as shown in the insert image from upper left to lower right in Fig. 8. These dots, as shown in the previous section, possess optical birefringence and are bright in transmitted light with crossed polarizers. Laser illumination of these modified regions for Raman analysis resulted in enhanced Rayleigh scattering of the laser beam, which is an indication of a difference in refractive index. The Raman spectra on each of the dots (in the end-on orientation) and on nearby, unmodified regions of the quartz glass were obtained (a total of 48 spectra). The spectra from the dots appear, on average, to have slightly less intense Raman feature than those from the unmodified areas. In addition, there appear to be subtle differences between the Raman bands from the dots and the unmodified regions over the frequency range (400-600 cm^{-1}) associated with small silica rings. A multivariate, chemometric analysis[25] on these spectra was performed to better distinguish these subtle differences. Using matrix algebra, this analysis simultaneously processed all 48 spectra in the set. This analysis looks for differences between spectra and generates a set of factors. These factors look like spectra and may be thought of as components of the experimental spectra. One important characteristic of the factors is that parameters used to scale the factors are called "scores". A score can be thought of as the relative contribution of a factor to one of the experimental spectra. Thus, the discrimination analysis generates one set of spectra-like factors for the entire set of experimental spectra but generates a separate set of scores (one for each factor) for each experimental spectrum.

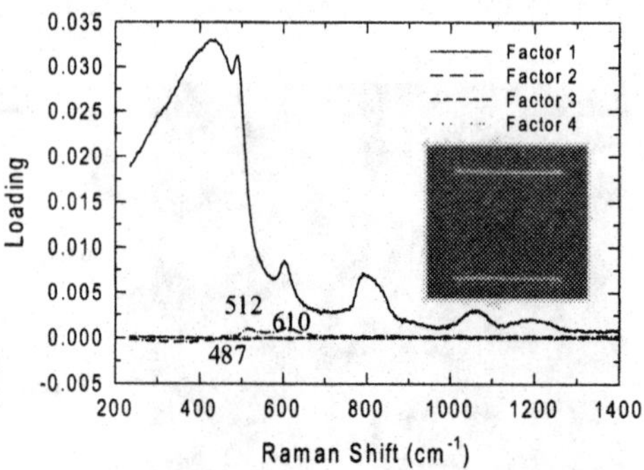

Fig. 8 Factors from discrimination analysis of Raman Spectra from laser modified and un modified regions of the square-dot silica glass (insert shows the square-dot pattern).

Fig. 8 plots the first four factors from the discrimination analysis of the 48 spectra. Factor 1 is the average spectrum of all the experimental spectra employed in the discrimination analysis. High-numbered factors are added to factor 1 to reproduce the experimental spectra. Factor 2 is significantly more intense than factor 3 and 4, which, in turns, are more intense than even higher-numbered factors. Factor 2, therefore, represents what is probably the most significant difference between the experimental spectra. Fig. 9 plots the scores for factors 1 and 2 from the discrimination analysis of Raman spectra from the laser modified and unmodified regions of the square-dot silica sample. Clustering of data points in Fig. 9 means that the spectra corresponding to the points in the cluster have similar contributions from factor 1 and factor 2. Separation of data points means the corresponding spectra have different contributions from factors 1 and 2. The data point for the unmodified regions are clustered at the lower right of the plot. In fact, all the unmodified (amorphous silica) data points have lower factor 2 scores than any of the modified data points. This indicates that the spectra from laser-modified regions have higher contributions from factor 2, at the expense of their

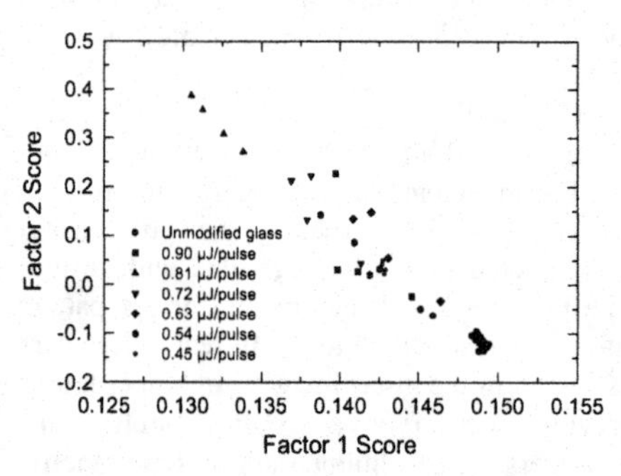

Fig. 9 A plot of scores of factor 2 versus the scores of factor 1. Each point corresponds to one of the spectra from modified and unmodified regions.

contribution from factor 1, than the spectra from the unmodified regions. Of the spectra from modified regions, those resulting from modification with a pulse energy of 0.81 μJ have the most factor 2, followed by the 0.72-μJ dots. The 0.90-μJ, 0.63-μJ, 0.54-μJ, and 0.45-μJ dots have less factor 2 than the 90-μJ and 80-μJ dots but more than the unmodified glass regions.

Factor 2 (Fig. 8) shows two prominent, positive features, one peaking at 512 cm^{-1} and one peaking at 610 cm^{-1}, plus a negative-pointing feature near 487 cm^{-1}. Positive-pointing features represent species that are more prevalent in spectra with higher factor two scores (like the modified regions) and vice-versa for the negative-pointing feature. The 610 cm^{-1} feature corresponds to 3-membered silica rings. The Raman feature for the 4-membered silica ring occurs near 490 cm^{-1}. The up-shift of the 4-membered silica ring to 512 cm^{-1} suggests that these 4-membered rings are under compressive stress. Therefore, the scores for factor 2 indicate enhanced concentrations in the laser modified regions with 3- and 4-membered silica rings, with some 4-membered rings created from unstrained precursors (see the negative-pointing 487 cm^{-1} feature in factor 2, Fig. 8). The increase of 3- and 4-membered silica rings by fs laser pulses has been reported in the literature.[14] These small rings may be forming at quasi-surfaces on small, perhaps nanoscale, voids regions associated with laser modification regions in glass.

Unlike factor 2, factor 3 and factor 4 do not discriminate between the spectra from the laser-modified and the unmodified regions. However, both the spectra form the 0.45-μJ dots and nearby unmodified regions show enhanced contribution from factor 4 (not shown). Factor 4 (when its plot is expanded to study it in detail) appears to correspond to variations in the amounts of 5-, 6- and 7-membered rings (the broad band peak near 450 cm^{-1}, Fig. 8). There is probably a small variation in bulk glass structure near where the 0.45-μJ modified regions were formed.

In summary, these Raman studies have revealed that intense fs laser pulses can reduce the population of large network ring structure in the glass and increase the number of 4- and 3-membered silica ring structures in glass. With these small network ring configurations, laser modified glass region can be more compact in comparison to the bulk glass. Furthermore, the observation of an up-shift of the Raman band for the 4-membered silica rings from 490 cm^{-1} to above 512 cm^{-1} suggests these rings are under compressive stress. [17, 26] These observations indicate local densification could contribute to the increase of refractive index change in the modified glass.

Laser-Induced Birefringence in Glass

The mechanism associated with the creation of birefringence in bulk glass using low repetition rate fs laser pulses with higher peak power energy is not clear yet. It is well known that glass can be poled under ultraviolet light,[27] or high dc bias condition at high temperature[28] to create a nonlinear optic behavior. A recent synchrotron study on the local atomic structural change by an in-situ laser irradiation has demonstrated a polarization-dependent structure change in amorphous chalcogenide glasses.[29] This incident light field or polarization dependent mechanism is especially applicable to fs laser processing since the estimated field generated by the laser pulse is on the order of 10^{11} V/m. The strong electrical field can effectively induced an anisotropic contraction by an electrostrictive effect.[30] Following this with an immense localized heating and rapid quenching, the overall effect could result in an irreversible electrostrictive deformation and create an anisotropic densification that is analogous to the mechanism proposed for ultraviolet-induced densification[27] of fused silica. Fig. 10 shows a microphotograph of a scanning electron microscopy (SEM) image of the photo-damaged region inside the glass through fracture. Fracture surfaces that exposed the damaged areas were obtained through a three-point bending experiment, where several photo-damaged waveguides written with pulse energy greater than 25 μJ were aligned at the center of the glass bend bar (where it had a maximum bending moment). A periodic nano-scale structure consisting of thin parallel glass plates and voids in the damaged areas was observed by SEM (see Fig. 10), where the laser-induced periodic structure is perpendicular to the laser polarization direction. The creation of this periodic structure perpendicular to the laser propagation direction suggests a possibly anisotropic contraction in the field direction. As this anisotropic contraction along the field direction increases with increased pulse energy, a periodic structure with a denser glass and void is created to preserve the conservation of mass. More interestingly, the periodicity of the laser induced nano-structure (~ 540 nm) is close to the wavelength of the incident laser (800 nm) traveling in silica (548.7 nm). Difficulties were experienced in identifying such period structures in laser modified region in glass when pulse energy is below the damage threshold (< 0.9 μJ/pulse). However, the observation of fs laser-induced nano-scale periodic structure in fused silica at 1 μJ/pulse was recently reported in the literature.[31] An increase in these nano-scale, periodic structures by fs laser pulses will increase the light scattering loss in waveguides, which is consistent with experimental observations presented in the previous section. It is believed that the laser-induced periodic structure with different refractive indices could also contribute to the development of optical birefringence in bulk glass. Birefringence of this nature is well known as "form" birefringence.[32]

Fig. 10 SEM microphotograph of photo-damaged region on a three-point bend fracture surface (arrow – laser beam direction)

SUMMARY

The important processing condition in creating embedded waveguides in a bulk glass by fs laser pulses was reported. Results indicate optical waveguides can be fabricated only within a very narrow pulse energy window (0.45 to 0.90 μJ/pulse). The direct fabrication of low loss, single mode waveguides, as well as some integrated optical devices by fs laser pulses has been demonstrated. These laser-modified regions possess birefringent behavior that can be controlled by the polarization direction of the incident laser beam. The phenomenon is explained in terms of a "form" birefringence, where a laser-induced periodic structure is created by an anisotropic contraction from the immense light field associated with fs laser pulses. The ability to create 3D low loss waveguides and birefringence properties in bulk glass by direct-write laser process will open up the untapped potential for fabrication of integrated 3D novel optical devices.

ACKNOWLEDGEMENT

The authors would like to acknowledge Gina Simpson for her technical support on Raman measurements. Sandia is a multiprogram laboratory operated by Sandia Corporation, a Lockheed Martin Company, for the United States Department of Energy under Contract No. DE-AC04-94AL85000.

REFERENCES

[1] P. Maine, D. Strickland, P. Bada, M. Pessot, and G. Mourou, "Generation of Ultrahigh Peak Power Pulses by Chirped Pulse Amplification," IEEE J. Quantum Electr. , **24** [2] 398-403 (1988).

[2] K. Yamasaki, S. Juodkazis, T. Lippert, M. Watanable, S. Matsuo, and H. Hisawa, "Dielectric Breakdwon of Rubber Materials by Femtosecond Irradiation," Appl. Phys., **A76** [3] 325-329 (2003).

[3] K. Miura, J. Qiu, H. Inouye, and T. Mitsuyu, "Photowritten Optical Waveguides in Various Glasses with Ultrashort Pulse Laser," Appl. Phys. Lett., **71**, 3329-3331 (1997)

[4] K. M. Davis, K. Miura, N. Sugimoto, and K. Hirao, "Writing Waveguides in Glass with a Femtosecond Lser," Opt. Lett., **21** [21] 1729-1731 (1996).

[5] D. Homoelle, S. Wielandy, A. L. Gaeta, N. F. Borrelli, C. Smith, "Infrared Photosenitivity in Silica Glasses Exposed to Femtosecond Laser Pulses," Opt. Lett., **24** [18] 1311-1313 (1999).

[6] J. D. Mills, P. G. Kazansky, E. Bricchi, and J. J. Baumberg, "Embedded Anisotropic Microreflectors by Femtosecond-Laser Nanomachining," Appl. Phys. Lett., **81** [2] 196-198 (2002).

[7] F. Garcia-Santamaria, C. Lopez, F. Meseguer, F. Lopez-Tejeira, J. Sachez-Dehesa, and H. T. Miyazaki, "Opal-Like Photonic Crystal with Diamond Lattice," Appl. Phys. Lett.,**79** [15] 2309-2311 (2001).

[8] M. Masuda, K. Sugioka, Y. Chen, N. Aoki, M. Kawachi, K. Shihoyama, K. Toyoda, H. Helvajian, and K. Midorikawa, "3-D Microstructure Inside Photosensitive Glass by Femtosecond Laser Excitation," Appl. Phys., **A76** 857-860 (2003).

[9] A. Marcinkevicious, S. Juodkazis, M. Watanable, M. Miwa, S. Matsuo, H. Misawa, and J. Nishii, "Femtosecond Laser-Assisted Three-Dimensional Microfabrication in Silica," Opt. Lett., 26 [5] 277-279 (2001).

[10] S. Cho, H. Kumagai, K. Midorikawa, and M. Obara, "Time-Resolved Dynamics of Plasma Self-Channeling and Bulk Modification in Silica Glasses Induced by a Hihg-Intensity Femtosecond Laser," *First International Symposium on Laser Precision Modification*, Editors: I. Miyamoto, K. Sugioka, and T. W. Sigmon, Proceeding of SPIE Vol. 4088, (SPIE, Bellingham, WA) 40-43 (2000).

[11] R. Osellame, S. Taccheo, M. Marangoni, R. Ramponi, P. Laporta, D. Polli, S. De Silvestri, and G. Gerullo, "Femtosecond Writing of Active Optical Waveguides with Astimatically Shaped Beams," J. Opt. Soc. Am. B, **20** [7] 1559-1567 (2003).

[12] Y. Chen, K. Sugioko, K. Midorikawa, M. Masuda, K. Toyoda, M. Kawachi, and K. Shihoyama, "Control of the Cross-section Shape of a Hollow Microchannel Embedded in Photostructurable Glass by use of a Femtosecond Laser," Opt. Lett., **28** [1] 55-57 (2003).

[13] C. B. Schaffer, J. F. Garcia, and E. Mazur, "Bulk Heating of Transparent Materials using a High-Repetition-Rate Femtosecond Laser," Appl. Phys., **A76** 351-354 (2003).

[14] P. Yang, D. R. Tallant, G. R. Burns, and M. L. Griffith, "Femtosecond Laser Pulse Induced Refractive Index Changes in Glass, " 2002 Material Research Society, "Femto- and Attosecond Phenomena in Material, Dec. 2-6, Boston, MA., (2002)., J. W. Chan, T. R. Huser, S. H. Risbud, and D. M. Krol, "Modification of the Fused Silica Glass Network Associated with Waveguide Fabrication Using Femtosecond Laser Pulses, Appl. Phys., **A76** 367-372 (2003).

[15] P. Yang, G. R. Burns, J. Guo, T. S. Luk, and A. G. Vawter, "Femtosecond Laser-Pulse-Induced Birefringence in Optically Isotropic Glass," to be appeared in the May issue of J. Appl. Phys., 95 [7] (2004).

[16] L. Sudrie, M. Franco, B. Prade, and A. Mysyrowicz, "Writing of Permanent Birefringent Microlayers in Bulk Fused Silica with Femtosecond Laser Pulses," Opt. Comm., **171** 279-284 (1999).

[17] P. Yang, G. R. Burns, D.R. Tallant, J. Guo, and T. S. Luk, "Direct-Write Waveguides and Structural Modification by Femtosecond Laser Pulses," *Micromaching and Microfabrication Process Technology IX*, Edited by M. A. Maher and J. F. Jakubczak, Proceeding of SPIE Vol. 5342 (SPIE, Bellingham, WA) 146-155 (2004).

[18] K. Hirao and K. Miura, "Writing Waveguides and Gratings in Silica and Related Materials by a Femtosecond Laser," J. Non-Crystallline Solids, **239** 91-95 (1998).

[19] K. Miura, J. Qiu, T. Mitsuyu, and K. Hirao, "Preparation and Optical Properties of Fluoride Glass Waveguides Induced by Laser Pulses," J. Non-Crystalline Solid, **256&257** 212-219 (1999).

[20] M. Will, S. Nolte, B. N. Chichkov, and A. Tunnermann, "Optical Properties of Waveguides Fabricated in Fused Silica by Femtosecond Laser Pulses," Appl. Opt., **41** [21] 4360-4364 (2002).

[21] B. J. Luff, R. D. Harris, and J. S. Wilkinson, R. Wilson and D. J. Schiffrin, "Integrated-Optical Directional Coupler Biosensor," Opt. Lett., **21** [8] 618-620 (1996).

[22] M. Hirano, K-I Kawamura, and H. Hosono, "Encoding of Holographic Grating and Periodic nano-Structure by Femtosecon Laser Pulse," Appl. Surface Sci., **197-198** 688-698 (2002).

[23] A. M. Streltsov and N. F. Borrelli, "Fabrication and Analysis of a Directional Coupler Written in Glass by Nanojoule Femtosecond Laser Pulses," Opt. Lett., **26** [1] 42-43 (2001).

[24] R. Gao, J. Zhang, L. Zhang, J. Sun, X. Kong, H. Song, J. Zheng, "Femtosecond Laser Induced Optical Waveguides and Micro-Mirrors Inside Glasses," Chin. Phys. Lett., **19** [10] 1424-1426 (2002).

[25] D. Haaland, K. Higgins, D. R. Tallant, "Multivariate Calibration of Carbon Raman Spectra for Quantitative Determination of Peak Temperature History," Vibrational Spectroscopy, **1** 35-40 (1990).

[26] D. R. Tallant, B. C. Bunker, C. J. Brinker, and C. A. Balfe, "Raman Spectra of Rings in Silicate Materials, " Mater. Res. Soc. Symp. Proc 73, 261-271, Material Research Society, Pittsburg, 1996.

[27] T. Fjjiwara, M. Takahashi, and A. J. Ikushima, "Second-Harmonic Generation in Germanosilicate Glass Poled with ArF Laser Irradiation," Appl. Phys. Lett., **71** 1032 (1997).

[28] W. Marguils, F. C. Garcia, E. N. Hering, L. C. Guedes Valente, B. Lesche, F. Laurell, and I. C. S. Carvalho, "Poled Glasses," MRS Bull., **23** 31-35 (1998), and references within.

[29] G. Chen, H. Jain, M. Vlcek, J. Li, D. A. Drabold, S. Khalid, and S. R. Elliott, "Study of Light-Induced Vector Change in the Local Atomic Structure of As-Se Glasses by EXAFS," J. Non-Crystalline Solids, **326&327** 257-262 (2003).

[30] N. F. Borrell, C. M. Smith, J. J. Price, and D. C. Allan, "Polarized Excimer Laser-Induced Birefringence in Silica," Appl. Phys. Lett., **80** [2] 219-221 (2002).

[31] Y. Shimotsuma, P. G. Kazansky, J. Qiu, and K. Hirao, "Self-Organized Nanogradings in Glass Irradiated by Ultrashort Light Pulses," Phy. Rev. Lett., **91** [24] 247405-1 (2003).

[32] M. Born and E. Wolf, *Principle of Optics*, Chapter 15, 835-840, Cambridge University Press, Cambridge, United Kingdom, 1999.

TUNABLE MICROPHOTONIC DEVICES IN FERROELECTRICS

David Scrymgeour
254 Materials Research Lab
University Park, PA 16802

Venkat Gopalan
253 Materials Research Lab
University Park, PA 16802

Kevin Gahagan
SP-FR-01-8
Corning, Incorporated
Corning, NY 14831

ABSTRACT:

This paper gives an introduction to the technology of microphotonic devices in ferroelectrics and will describe both the theory of operation and design considerations. The fabrication process for several microphotonic devices created in a single ferroelectric chip of $LiNbO_3$ and $LiTaO_3$ will be introduced. The performance of these devices will then be described, such as dynamic focusing lens stack of microlenses whose focal power can be tuned by electric field and an optical scanner that can steer laser light with application of an electric field. A first-of-its kind, integrated scanner and lens device will also be described. Emerging technologies based on microphotonic devices in ferroelectrics will also be explored.

INTRODUCTION:

The ability to control the angular position and the spot size of a laser beam with high speed are of interest in many applications including optical communications, optical data storage, laser printing, analog to digital conversion, and display technologies. Of the many competing technologies, solid-state electro-optic devices based on ferroelectrics such as lithium niobate ($LiNbO_3$) and lithium tantalate ($LiTaO_3$), have several advantages over mechanical and other systems including the absence of moving parts, small device sizes, and high operating speeds (intrinsic response speeds electro-optic effect is in the gigahertz range).[1-6] In addition, devices based in these materials can provide a versatile solid-state platform for microphotonics where such diverse optical functions such as optical scanning, shaping, focusing, and frequency conversion, can all be seamlessly integrated on the same chip.

The electro-optic effect is the change of the index of refraction of a material with an applied electric field. The electro-optic effect, also called the electric-field tunable refractive index, is currently of great interest for light modulation at speeds of tens of Gigahertz. In a crystal, the electro-optic effect is given as

$$\Delta n_e = -\frac{1}{2} n_e^3 r_{33} E_3 \tag{1}$$

where n_e is the extraordinary index of refraction, E_3 is the applied electric field, r_{33} is the electro-optic coefficient, and Δn_e is the change in the index of refraction. For polarized light traveling in a such a crystal the index of refraction, n_e, which determines the speed of the light through the material, will change to $n_e + \Delta n_e$ when an electric field is applied to the material.

Two very important materials that posses the electro-optic effect along with excellent optical properties are lithium niobate ($LiNbO_3$) and lithium tantalate ($LiTaO_3$). They are both *ferroelectric* materials, which are a class of materials that posses spontaneous polarization of the lattice which can be realigned with the application of an external electric field. Regions of uniform polarization within a material are called *domains*, and the boundaries between two different domains are called *domain walls*. Both $LiTaO_3$ and $LiNbO_3$ have two allowed domain orientations, one pointing in the positive direction along the z crystallographic axis (up domain), or oriented 180° to this direction pointing along the –z crystallographic axis (down domain). The sign of the electro-optic coefficient, r_{ij}, depends upon the orientation of these domains, so light shining through the crystal under an uniform electric field will see a decrease in the index of refraction ($n - \Delta n$) in one domain area and a increase ($n + \Delta n$) in the other domain orientation, and a change index of $2\Delta n$ at the domain wall. By patterning domain in these materials into particular shapes, active deflection and shaping of laser beams can be achieved.

DESIGN OF DEVICES

Scanners

The tunable index change present at domain walls can be exploited to deflect light and create optical scanners by patterning the domains into a series of prisms. First proposed by Lotspeich in 1968, initial scanner designs consisted of N identical prisms placed in sequence, each with a base l and height W, such that the total length of the rectangular scanner is $L=Nl$ and width is W.[3] The total deflection angle, θ_{int}, at the output for a light beam incident along the axis, L, of the so-called rectangular scanner is given by

$$\theta_{int} = n_e^2 r_{33} E \frac{L}{W} \tag{2}$$

there n_e is the index of refraction of the material, and r_{33} is the electro-optic coefficient, and E is applied electric field. This results in an electric field controlled deflection angle θ_{int}. When the deflected light beam exits the electro-optic material (index, n) into air (index, 1), the output deflection is enhanced further by Snell's law as $\theta_{ext} = \sin^{-1}(n \sin(\theta_{int}))$. A beam propagation method (BPM) simulation of such a scanner is shown in Figure 1(a). For maximum deflection, the beam must be contained in the stack, so this sets the limitation for increasing the angle by decreasing the width and increasing the length simultaneously.[6, 7]

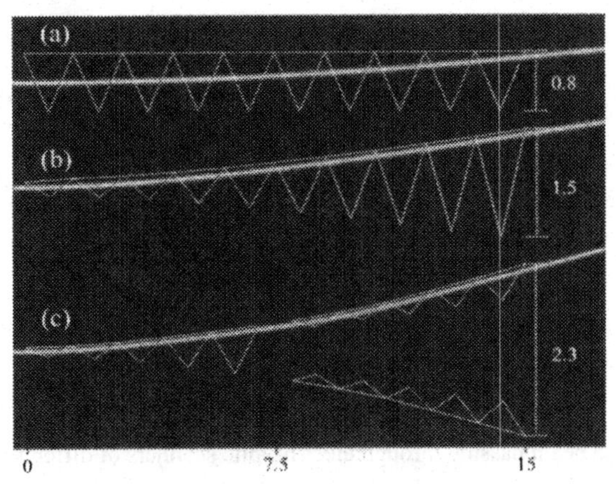

Figure 1: Beam propagation method simulation of different electro-optic scanner designs in $LiTaO_3$: (a) rectangular, (b) horn shaped, and (b) cascaded horn shaped. Deflection angles (one way) are 5.25°, 8.45°, and 14.77° respectively. All dimensions are in mm. Each device has the same length (15 mm), operating field (15 kV/mm), interface number (10), and beam size (100 μm). Domain orientation inside the triangles is 180° opposite to that of the surround areas.

An improvement of this design allows the width of the scanner to stay as small as permitted by beam diameter at the input, but gradually increased to just accommodate the trajectory of the beam. This is the so-called *horn shaped scanner*, as the width of the scanner flairs toward the end of the device to accommodate the additional beam deflection.[8] A schematic is shown in Figure 1(b) and has been demonstrated in $LiTaO_3$.[5,9,10]

Recently, a new design concept called *cascaded horn shaped scanner* has been proposed.[11] It consists of a series of individual scanners aligned so that subsequent scanners in the stack are aligned to the previous scanner's peak deflection. A schematic of a 2 stage cascaded scanner is shown in Figure 1(c). At peak operating field, the first scanner first deflects the beam through a given angle and deflects along a given direction. The second scanner is oriented along this direction, so that when a field is applied to the second scanner further deflection occurs. All scanners would be fabricated in the same piece of electro-optic crystal, with each scanner in the stack having a separate electric field supply - in effect cascading the scanners. The advantage this design is that additional stages can be appended until a final target deflection is achieved, or the field requirements can be reduced for a given required scan angle (by increasing the number of stages). The trade off is in more complex drive controllers, as each scanner must work in tandem with the others.

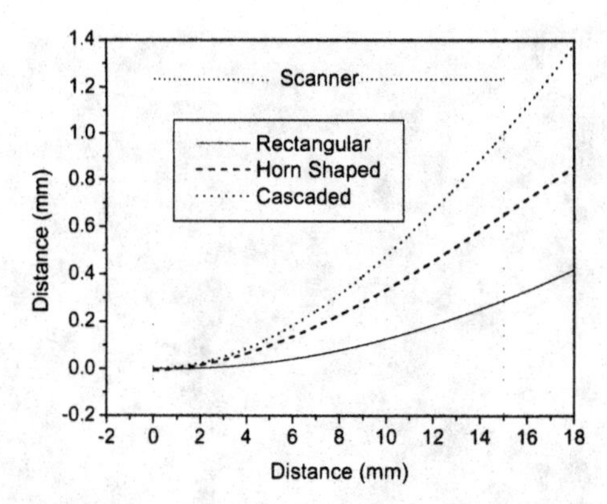

Figure 2 : Path of the beam passing through electro-optic scanners of different design type shown in Figure 1.

A comparison of all these scanner designs is shown quantitatively in Figure 1 where the input aperture, applied field, and number of interfaces are the same for the rectangular, horn-shaped, and cascaded horn shaped scanners. Figure 2 shows the deflected beam paths as function of distance passing through each scanner. It is evident that the horn-shaped scanner offers much improved performance, and the cascaded horn-shaped scanner even better performance.

Lenses

Similar to the scanners in a domain reversed ferroelectric material, patterning domains in the shape of lenses can create lens stacks to focus and diverge beams passing through the crystal. If the domain is patterned into a stack of simple thin lenses, the power, ϕ, of the combined lens is given by

$$\phi = \frac{1}{f} = N\left(\frac{n_2 - n_1}{n_1}\right)\left(\frac{1}{R_2} - \frac{1}{R_1}\right) = N\left(\frac{2\Delta n_e}{(n_e - \Delta n_e)}\right)\left(\frac{1}{R_2} - \frac{1}{R_1}\right) \quad (3)$$

where N is the number of lenses, f is the focal length, R_1 and R_2 are the radii of curvature of the front and back face, respectively, and the indices inside and outside the lenses are given by $n_1 = (n_e - \Delta n_e)$ and n_2 is $(n_e + \Delta n_e)$, respectively.

Shown in Figure 3 is a beam propagation method simulation of a tunable focusing lens stack. It is designed to focus a highly diverging beam input, for example the output of a fiber optic cable, and to focus the beam to a collimated output. Many lenses are needed because the index contrast is small.

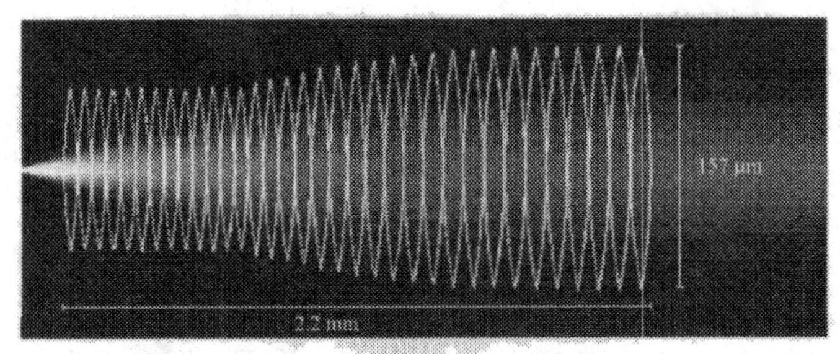

Figure 3: A BPM propagation simulation of a 32 biconvex lens stack. Each lens is spaced 5 µm apart. The stack is designed to collimate a diverging input beam of 4 µm to 100 µm at the output with an applied field of 8 kV/mm.

FABRICATION

The key to fabrication of devices is the creation and shaping of domain areas in a single crystal, called domain micropatterning. In this process, commercially obtained crystals 0.3 mm thick are diced pieces about 20 x 20 mm. Using a lithographic process, a tantalum metal pattern is created on the surface of the crystal resembling the desired device, for example the horn-shaped scanner shown in Figure 1, with the metal defining the complement of the area of the prisms (or lenses) you wish to create. A uniform water electrode is placed in contact with the bottom crystal surface and a large external electric field greater than the coercive field of the material (21 kV/mm or ~6000 V across the 0.3 mm crystal) is applied to the patterned tantalum film which reverses the domain orientation in areas between the top metal electrode and the bottom water electrode.

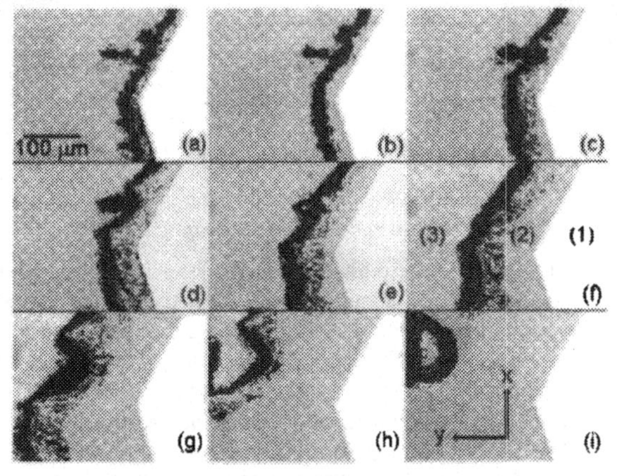

Figure 4: Selected frames from *in situ* observation of domain growth in patterned LiTaO3. Three regions, labeled in frame (f) are: the original crystal beneath the Ta film electrode [region (3)], the domain-inverted region underneath the Ta-film electrode [region (2)], and the original crystal with no Ta-film electrode forming the prism pattern [region (1)]. The dark region is contrast at the boundary between domains. Domain growth starts at the electrode edge and advances into the electrode. Each frame is separated by 3 seconds.

Precise control of the domain movement is achieved by monitoring the external currents flowing in the circuit as well as directly observing the domain nucleation and growth *in-situ* by monitoring using electro-optic imaging microscopy (EOIM). This process takes advantage of the index difference $2\Delta n_o$ across a domain wall on application of an electric field - the same concept used for operation of the electro-optic scanner. This index difference at the wall will cause scattering of transmitted or reflected light in the crystal. This can be used to image domains walls through the crystal thickness in a z-cut crystal in an optical microscope, with or without any polarizers. In this way, the movement of the 180° domains could be directly observed during the domain patterning process.[10, 12] Figure 3 shows selected frames from the in-situ observation of such a poling process during the device fabrication. The spacing between successive frames is 3 seconds. The white triangular shaped area at the right of each frame is the vertex of one of the prism triangles that does not have Ta-film, while the rest of each frame is the Ta electrode. Nucleation occurs at the electrode edges and advances to the left into the Ta electrode.

TESTING:

After domain patterning, the input and output faces are polished to an optical grade using diamond suspensions. Uniform metal electrodes are then sputtered on each side of the device and connected to copper tape that forms the leads of the device. The entire device with the exception of the input and output faces is then encapsulated in silicon gel to inhibit breakdown during device operation.

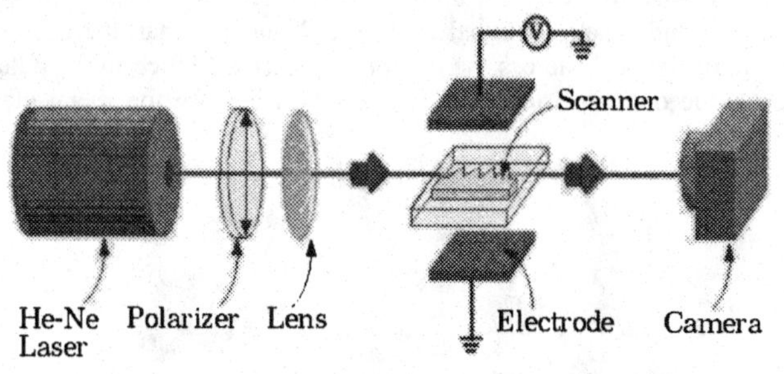

Figure 5: Device testing setup.

The device is then placed in the path of a He-Ne laser with a wavelength of 632.8 nm. Using a lens the beam is focused to cleanly pass through the device and shine onto a CCD camera. Applying different biases to the device will deflect the beam in the plane of the device for a scanner, or change the focal length of the beam for a lens. By measuring the deflection or spot size of the beam with the CCD camera, precise angular displacement information can be obtained. To test the lenses, both the position of the device relative to the focal point of the beam as well as the CCD camera position relative to the output face of the crystal is varied. By measuring the beam waists for a variety of positions, an analytical expression can be fit to the data.

DEVICE PERFORMANCE:

Several device designs have been fabricated and tested.[5,10,11] The first device to be discussed is an integrated lens and scanner devices pictured in Figure 6(a). The lens stack, which is the small rectangular section on the left side of the device in Figure 6(a), consists of 32 bi-convex lenses and is simulated in Figure 3. It is designed to focus a highly divergent beam, like the output from a fiber optic cable, shining on the input face and make it a collimated beam which then passes into the scanner portion which is the large rectangular area of the device on the right of Figure 6(a). The performance details are shown in Figure 6(b),(c). In (b) notice for a fixed position along the x axis, that for different apply voltages one can get different beam sizes. The displacement in (c) is linear with applied voltage and in excellent agreement with simulations. For polarized input light, the measured voltage dependence of the deflection angle is 31.45 mrad/kV (1.802°/kV) and the maximum deflection angle is ±7.44° (a total of 14.88°) at ±4.05 kV. A theoretical estimate as determined by BPM simulation is plotted on the same graph (see Figure 6). Simulation gives a voltage dependence of the deflection angle as 31.41 mrad/kV (1.80°/kV) with a maximum deflection of ±7.295° (total 14.59°) at ±4.05 kV.

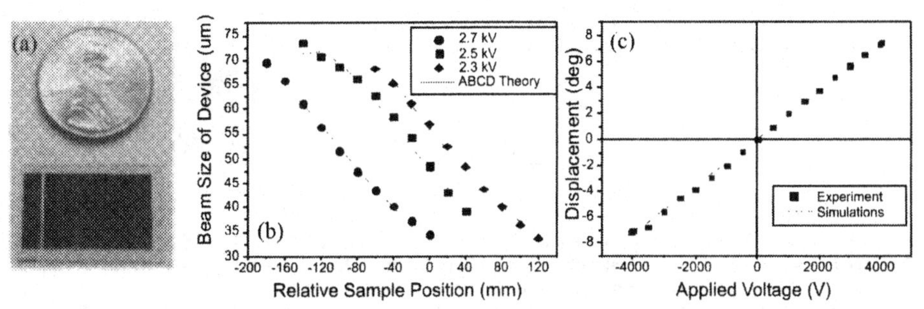

Figure 6: (a) Integrated lens and scanner device next to a penny. Scanner section small rectangle on left, scanner is rectangle on right. The performance of the lens stack is shown in (b), which shows the beam waist for different voltages as a function of different position of the device relative to the focal point of the laser along with fits from ABCD theory. (c) shows the deflection of the beam for different applied voltages along with a best fit line from the BPM simulations. Note the linear response.

A cascaded electro-optic scanner has been fabricated and tested and is pictured in Figure 7 . Figure 8(a) shows how the first and second stage scanners would work in tandem. Peak deflection is reached when both the first and the second scanners are at peak voltages. For continuous scanning, one needs a voltage driver that seamlessly applies the appropriate voltage waveforms to both stages. The scanner was designed to work in conjunction with a high-speed voltage supply with peak output voltages of 1.1 kV. A specially designed arbitrary waveform signal generation based in a field programmable gate array (FPGA) which outputs to high voltage amplifiers was built in collaboration with the University of Delaware.[13]

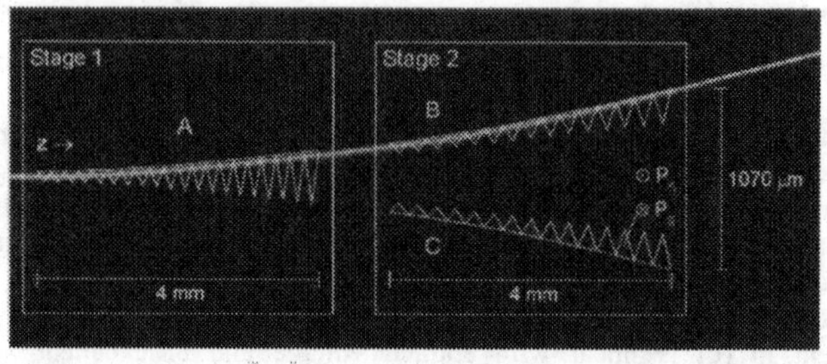

Figure 7: A BPM simulation of the cascaded electro-optic scanner showing peak deflection from both scanners at 11 kV/mm.

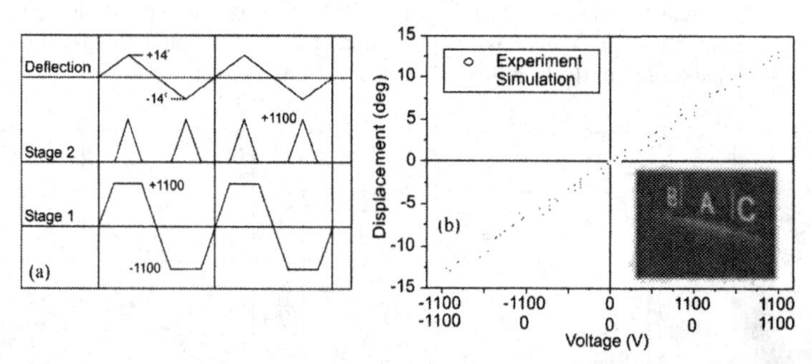

Figure 8: (a) The driving waveforms for optimal operation. The bottom trace shows the voltage placed on the stage 1 (scanner A). While at peak voltage, the voltage for stage 2, scanners B,C (middle trace) is ramped. The top trace shows the deflection of the beam. (b) Performance of high angle scanner. The upper row of voltages refers to the first stage voltage and lower row refers to second stage voltage. The inset shows the high-speed scan (5 kHz) showing full 25.4°. The beam is projected on a flat screen and imaged from the side. The two vertical lines indicate different scan areas.

The deflection angle is measured as a function of applied voltage and is plotted in Figure 8(b). For extraordinary polarized input light (along the ferroelectric polarization direction), the measured voltage dependence of the deflection angle is 102.0 mrad/kV for scanner 1 and 99.5 mrad/kV for scanner 2 and the maximum deflection angle for the device is ±12.7° (a total of 25.4°) with ±1.1 kV on both scanner stages. A theoretical estimate as determined by the BPM simulation is plotted on the same graph. The simulation predicts a voltage dependence of the deflection angle is 104.7 mrad/kV for scanner 1 and 102.2 mrad/kV for scanner 2 with a maximum total deflection of ±13.04° (total 26.8°) for peak fields on both scanners. The maximum number of resolvable spots for this device was estimated as 29.75 spots at maximum total deflection. Finally, the scanner operation was also tested up to 5 kHz. A picture of the scanned beam projected on a screen is shown in the inset of Figure 8(b). No noticeable degradation of the scan angle was observed, and higher scan speeds should be possible with higher power drivers.

CONCLUSIONS

The design and fabrication of several devices have been performed including the fabrication of an integrated lens and scanner device and a new cascaded electro-optic device which shows the highest scan angle so far reported for this technology of 25.4° at 5kHz. The further development of these devices can be used in next-generation optical devices and systems, including space based communication systems where non-inertial beam deflection is required. Integrating these devices with other structures currently fabricated in ferroelectrics, like second-harmonic generation structures, can allow for many an exciting way to integrate several important optical functions, including beam deflection, beam shaping, and frequency doubling, all seamlessly integrated on the same optical chip.

Acknowledgements:

This work was performed under the supervision of Dr. Venkatraman Gopalan at Penn State University. The author would like to gratefully acknowledge the support of the DARPA-STAB program, the Los Alamos National Laboratory, and the National Science Foundation grant (ECS 9988685).

References:

[1] Y. Ninomiya, NHK Technical Journal **26**, 1 (1974).
[2] Y. Ninomiya, Electronics and Communications in Japan **56**, 108 (1973).
[3] J. F. Lotspeich, IEEE Spectrum, 45 (1968).
[4] V. Gopalan, K. T. Gahagan, M. Kawas, et al., **25**, 371 (1999).
[5] K. T. Gahagan, V. Gopalan, J. M. Robinson, et al., **38**, 1186 (1999).
[6] Q. Chen, Y. Chiu, D. N. Lambeth, et al., Journal of Lightwave Technology **12**, 1401 (1994).
[7] K. T. Gahagan, V. Gopalan, J. M. Robinson, et al., Proceedings of the SPIE - The International Society for Optical Engineering **3620**, 374 (1999).
[8] Y. Chiu, J. Zou, D. D. Stancil, et al., Journal of Lightwave Technology **17**, 108 (1999).
[9] J. C. Fang, M. J. Kawas, J. Zou, et al., IEEE Photonics Technology Letters **11**, 66 (1999).
[10] D. A. Scrymgeour, Y. Barad, V. Gopalan, et al., **40**, 6236 (2001).
[11] D. A. Scrymgeour, A. Sharan, V. Gopalan, et al., Applied Physics Letters **81**, 3140 (2002).
[12] V. Gopalan, Q. X. Jia, and T. E. Mitchell, Applied Physics Letters **75**, 2482 (1999).
[13] F. Muhammad, P. Chandramani, J. Ekman, et al., IEEE Photonics Technology Letters **14**, 1605 (2002).

SOL-GEL PROCESSING OF $BaTiO_3$ FOR ELECTRO-OPTIC WAVEGUIDE DEVICES

D Bowman and S Bhandarkar
New York State College of Ceramics at Alfred University
G Kowach
City College of New York

ABSTRACT

The effects of processing parameters on the optical properties of barium titanate thin films were studied. Barium titanate is a good candidate for electro-optic devices such as a Mach-Zehnder waveguide modulator due to its high electro-optic coefficient. Thin films prepared by sol-gel deposition were optically characterized using a prism coupler. The sol was prepared with a large polymeric additive in an attempt to attain thicker sintered films. The waveguide loss of the film was shown to increase sharply at the onset of organic burn-off of this additive. It is postulated that an optically active 'microstructure' results from this burn-off that causes high loss due to light scattering. This domain-based structure also appears to template the growth of nanosized grains during the final sintering of the film.

INTRODUCTION

Over the past decade there has been a dramatic increase in the volume and speed of optical communications in the United States. In order to accommodate this increasing trend, there will be a continuing need for ultra fast as well as more efficient optical components. In this work, we focus on an enhanced version of the signal generation system used in optical networking. This is the one of the key components of the transmission link. Broadly speaking, the light source may be modulated directly (by turning the laser on and off to generate bits) or externally. Externally modulated sources use a continuous wave (CW) laser and use an external optical device to convert the digital electrical signal to an optical one. One such device is an electro-absorption modulator that uses the Franz-Keldysh effect to transmit or block light based on an external electrical stimulus.

The preferred device tends to be an electro-optic Mach-Zehnder waveguide modulator. The incoming light is broken into two pathways using a Y junction or a 3dB coupler. At the other end of the device, the arms recombine. If a characteristic voltage V_π were to be placed across one of the two arms as the light passes through it, the resultant π phase shift would cause destructive interference and we get an "off" output signal. Actual modulators may use bias so as to operate at the point of quadrature. The substrate is a Lithium Niobate ($LiNbO_3$) single crystal with a strip of titanium diffused into it to create the required change in index to confine light. This is typically done photolithographically. Electrodes are then evaporated onto the surface, leads attached, fibers coupled onto the ends and the device packaged appropriately. Various designs such as push-pull or dual-drive configurations have been implemented to optimize the system performance. Additionally today's modulators are designed to be traveling-wave modulators where the phase of the optical and the electrical wave is matched by making the effective index of propagation for each the same. These types of devices are commercially available for operation up to 40Gbps.

There are several ways we would like to enhance this device using integrated optics. Here we seek to investigate the prospects of depositing an electro-optic film onto a silica-clad silicon wafer. We would like this film to have thickness in the micron range to allow easier coupling to the laser or fiber light source. We seek a material with a high electro-optic coefficient so as to minimize the driving voltage, V_{π}, and a high switching speed so that the system will have the potential of exceeding 40Gbps. The salient feature of advanced integrated optical systems is the idea that one can combine active components such as these modulators with other passive components on a single chip. With continued development of glass-based solid-state lasers (DFB/DBR designs), it also leads to the possibility of combining different types of actives on the same substrate. Silicon is the preferred substrate certainly for electronics and increasingly for planar lightwave circuits (PLCs) and hence is the substrate used here.

This drive toward miniaturization and integration has been the goal of much research in the past years. Recently, perovskite ceramics such as PLZT and Barium Titanate have been shown to grow good films on silica on silicon and MgO substrate[1,2]. Other common ferroelectric materials have been studied for their suitability as waveguiding materials as well[3,4]. Several different deposition methods have been used to fabricate these films. The majority of them are vacuum vapor-phase deposition techniques including Molecular Beam Epitaxy, RF Magnetron Sputtering, Pulsed Laser Deposition and Metal-Oxide Chemical Vapor Deposition. Several groups working with these techniques have deposited films on MgO substrates that are capable of guiding and switching light with relatively low and acceptable losses [5,6]. Erbium, the amplification ion of choice for the C-band telecom window, has also been incorporated into their films and in at least one instance and an appreciable gain in signal strength has been demonstrated[7].

In general though, there is a great disparity in the propagation losses reported in the literature for different techniques and different groups. A sampling of the literature on waveguide loss in electro-optic thin film systems is shown in Table I. Loss measurements are not discussed in detail in several of these references and spurious effects can occur with high index contrast waveguide configurations. It is possible for cladding modes to propagate through the substrate thereby leading to spurious loss results for the slab. Typical prism-coupling techniques are particularly prone to this artifact. There appears to be little understanding as to what exact mechanisms are responsible for this large spread of the propagation loss.

In this work, we attempt to elucidate this issue and eventually optimize the propagation loss using a simplified deposition technique, Chemical Solution Deposition. The material we have chosen to focus on is barium titanate (BT), since this is a chemically easier system to investigate over others such as PLZT. Additionally, electro-optic coefficients for BT are shown to be among the highest for all ferroelectrics[8].

Table I: Optical clarity of polycrystalline electro-optic films

Reference	Deposition Method	Material	Substrate	Loss (dB/cm)	Δn
Adachi et al[17]	Sputtering	PLT	R-Sapphire	10	0.7
Petraru et al[18]	Pulsed Laser Dep	BT	MgO	1-2	0.6
Nashimoto et al[19]	Solid Phase Epitaxy	PZT	$SrTiO_3$	8	0.03
Nashimoto et al[20]	Solid Phase Epitaxy	PZT	Nb-$SrTiO_3$	1	0.03
Uhlmann et al[21]	Sol-Gel	PLT/PZT	Glass	1.4	0.75
Busch et al[22]	Sol-Gel	PZT	$SrTiO_3$	6-7	0.03
Urlacher et al[23]	Sol-Gel	PT (amor)	Pyrex	1	~0.3
Jin et al[24]	Sol-Gel	PLZT	R-Sapphire	2.7	0.7
Fuflygin et al[25]	Sol-Gel	$BaPbTiO_3$	R-Sapphire	2.5	0.7
Jiwei et al[26]	Sol-Gel	PZT	$SrTiO_3$	14	0.03

There are some key benefits to using sol-gel as a processing method. Firstly, the processing of the films is somewhat simplified as there are fewer variables to optimize. Most vapor-phase vacuum processes need complex equipment with multiple interacting systems each capable of influencing the properties of the film. The second key benefit is the ability to closely control the composition of a multi-component material. Differences in volatility of individual constituents of a multi-component film can make vapor-phase processing more challenging. Using a wet-chemical method, the cations are complexed through organic moieties and their ratios can be precisely controlled. Additionally, sol-gel is not a capital intensive technique.

However, the method is critically dependent of the subtleties of the chemistry of the sol. There are many recipes reported in the literature with the typical differences being the ligand and/or solvent used. The second problem is the thickness of the average film. Most sol-gel recipes require many, often more than 10 iterations of the deposition-pyrolysis process. Typical thickness of a single fired film will be in the order of 100 nanometers and this is unsuitable. We would like to deposit a film of micron-range thickness in a single casting and heat treatment. The third consideration is scattering losses in the polycrystalline film, which is what this study found to be the greatest problem.

After reviewing the literature on barium titanate sol-gel, a recipe designed by Kozuka, et al. was chosen for study[9]. It uses an acetate based sol that includes a large polymeric additive, polyvinylpyrrolidone (MW=630,000). Their study of this system has resulted in a good understanding of the gel-to-film conversion process. The PVP polymer has been shown to mitigate cracking in BT films so that these films can be made as thick as 1500 nm from a single coat[10]. They do not however report on the light-guiding properties of the film. The goal of this research is to take this work and understand how processing effects the microstructure and in turn the waveguide properties.

EXPERIMENTAL

An acetate-based barium titanate sol was prepared using the method described by Kozuka et al.[11] Reagents used were barium acetate (99+% $Ba(CH_3COO)_2$; Aldrich), titanium ethoxide ($Ti(OC_2H_5)_4$; Aldrich), poly-vinylpyrrolidone (C_6H_9ON, MW = 630,000; Alfa Aesar), distilled water (18MΩ), glacial acetic acid (99.9% CH_3COOH; Fisher) and reagent-grade ethanol (~90% C_2H_5OH; Fisher). Components were batched for a 1/1/0.5/5/30/5 mole ratio respectively (where the monomer of poly-vinylpyrrolidone (PVP) was used in the calculations). First, PVP was dissolved in a solution of ethanol and two thirds of the total volume of acetic acid. This process took several minutes of constant stirring. Titanium ethoxide was added drop wise once the PVP had been completely incorporated. The barium acetate was dissolved in a separate solution of water and the remaining volume of acetic acid. This process also took several minutes of stirring. The barium solution was added drop-wise to the titanium solution and let stir until a transparent sol was achieved. Typical batch size was 100 grams and was of sufficient volume to coat approximately 20 6" wafers.

Before coating, each wafer was cleaned with isopropyl alcohol and analyzed using the prism coupler. Cladding thickness and index was measured and logged for future computations. In all cases, the thickness of the silica cladding that was deposited via a flame hydrolysis procedure separately was greater than 15 micrometers and sufficiently thick to mask the effects of the substrate. Spin coating was performed using a Model G3-8 Desktop Precision Spin Coating System (Cookson Electronics). Approximately 2cc of barium titanium alkoxide sol was dispensed onto the wafer using a small syringe equipped with a 0.22μm nylon mesh filter. The samples were spun coat at speeds from 500 to 2000 rpm for approximately 30 seconds. After spin coating was complete, the wafer was transferred to a hot plate and dried at 100°C. The wafer was let cool and cleaved into samples of varying geometry for testing.

Thermal analysis of the sol was performed using an SDT 2960 Simultaneous DSC-TGA (TA Instruments). Thermo-Gravimetric Analysis (TGA) and Differential Scanning Calorimetry (DSC) were run simultaneously on the sample. Weight loss of the sol and heat flow of the furnace was monitored as the sample was heated in air from room temperature to 700°C at 10°/min in a platinum tray. Thermal Advantage software (v 1.1A, TA Instruments) was used to collect data, which was then analyzed using Universal Advantage software (v 3.4C, TA Instruments). Diffraction experiments were run on a Siemen's Kristalloflex diffractometer using Copper Kα x-rays and a Bragg-Bentano Goniometer. Small wafer samples were cleaved and mounted in a side-drift sample holder using X-ray amorphous putty. Data was collected from 10° to 65° 2θ in 0.02° steps, 10 seconds each. Datascan software (v 3.1, Materials Data Inc.) was used to collect intensity data from the diffractometer and Jade software (v 6.0.3, Materials Data Inc.) was used for analysis.

Sample microstructure was observed using scanning electron microscopy. A Philips 515 SEM was used in both secondary electron and backscattered electron modes to image the thin film. Small samples were cleaved, mounted onto aluminum posts using conductive epoxy and sputter-coated with approximately 100 Angstroms of 60Au-40Pd metal. A spot size of 100nm was used under 20 keV excitation. EVEX software was used to collect images of the microstructure. Optical imaging was performed using a Reichert-Jung Polyvar MET optical microscope operating in reflected light mode. Sections of samples were cleaved and observed under 20x to 1000x magnification. The surface roughness of the samples was probed using a light interference method. A Zygo Interferometer was used to model surface contours and statistically derived roughness information. Linux-based software was used to analyze the tens of thousands of data points collected. The primary optical characterization technique used for this study was prism coupling. This method allows the measurement of film thickness, index of refraction and waveguide loss, the three properties of greatest interest to us. A Metricon 2010 Prism Coupler using a HeNe laser (633nm) and a high index prism (~2.00) was used to couple light into the samples for analysis.

RESULTS AND DISCUSSION

The as deposited films were colorless both as spun, and after drying on the hot plate. The initial experiments examined the effect of firing the films at different rates. The thermal cycle described by Kozuka, et al.[12] was used for the preliminary tests. Samples were heated stepwise to 300°C to drive off volatile species, to 500°C to complete the combustion of the organics and then 700°C to sinter the ceramic. The results showed that by directly inserting the sample at the desired temperature (a process similar to rapid thermal annealing or RTA), finer grains sizes and lower losses could be realized. This effect has been demonstrated in the literature using PLZT films as well[13]. Figure 1 shows X-ray diffraction data taken from these samples. The [110] peak of these patterns was modeled and the breadth was used to calculate crystal size using the Scherer equation. The results of these calculations are given in Table II. Evidently, the samples directly inserted into the furnace show the smallest crystal size.

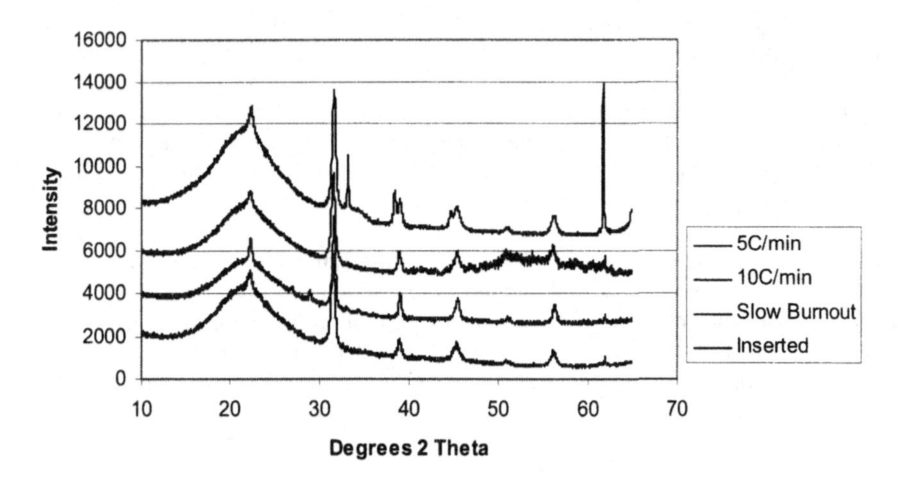

Figure 1: X-ray Diffraction of Samples Fired at Different Rates

Table II: Crystallite Sizes of Samples Fired at Different Rates

5°C/min to 300, 500, 700°C	196Å
10°C/min to 300, 500, 700°C	181Å
5°C/min to 300, 700°C, ½°C/min to 500°C	260Å
Direct Insertion to 300, 500, 700°C	173Å

These first samples were visibly scattering and prism coupling data corroborated this observation. A sample spectrum is shown in figure 2. The width of the transmission mode is a fair indicator of the lossy behavior of the material. Here, the band is seen to be very broad, indicative of high losses. Since only a single mode was detected, it was not possible to generate thickness of index data from this measurement.

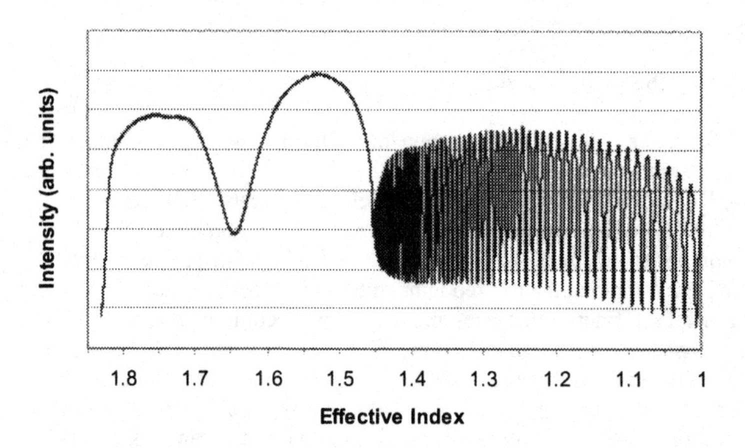

Figure 2: Prism Coupling Spectrum of Sample Fired at 5°C/min

Films were deposited by spin coating at several speeds and the dried gel film was measured using the prism coupler. The results of this test are shown in figure 3. In the region tested, spin speed can be related to the thickness of the dried gel film using a linear fit. Since these films were unfired, the index was consistently low, around 1.56. The propagation modes were very narrow as shown in figure 4, which indicates a low loss, approximately 1dB/cm loss, which is remarkably low in comparison to the fired films discussed later.

We then investigated the propagation loss in the films changed as the processing temperature was increased. Several samples were inserted into a furnace at high temperatures in ambient atmospheres and then tested for propagation loss using the prism coupler. The thickness of the dried gel was a limiting factor since films made using spin-speeds greater than 2000 rpm cracked upon firing. The cracking of these thicker films in our case is curiously different from the results from Kozuka et al[10].

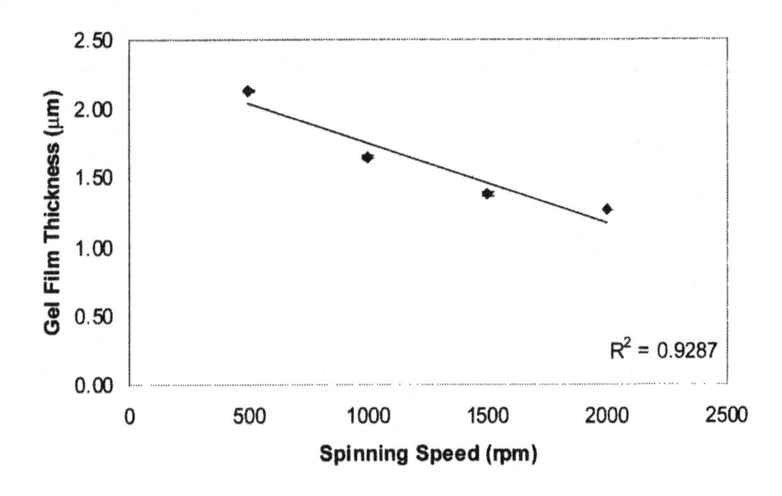

Figure 3: Dried Gel Film Thickness as a function of Spin Speed

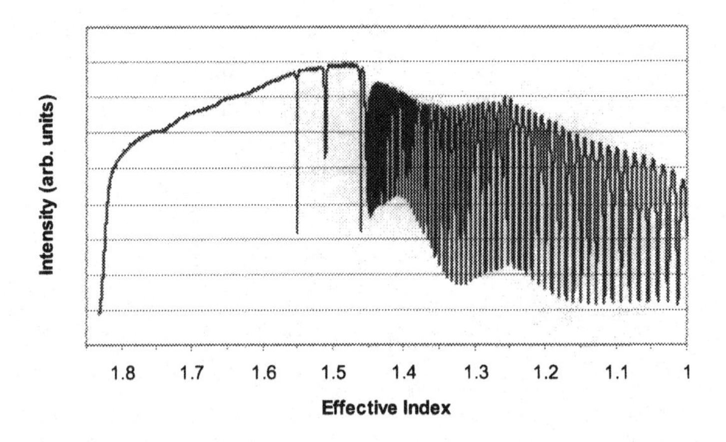

Figure 4: Prism Coupling Spectrum of Dried Gel Film

As seen in figure 5, at approximately 250ºC the loss of the samples increased sharply from 1-2 dB/cm to in excess of 15dB/cm. Data collected at higher temperatures was not meaningful as the maximum loss capable of measurement with a prism coupler is approximately 15dB/cm, corresponding to ~95% loss. The stock sol was analyzed using a simultaneous DSC-TGA, shown in figure 6. At approximately 250ºC, there is a loss in mass accompanied by an exothermic peak. It has been reported that this peak is associated with the burn-off of the additive poly-vinylpyrrolidone, using FTIR analysis[11].

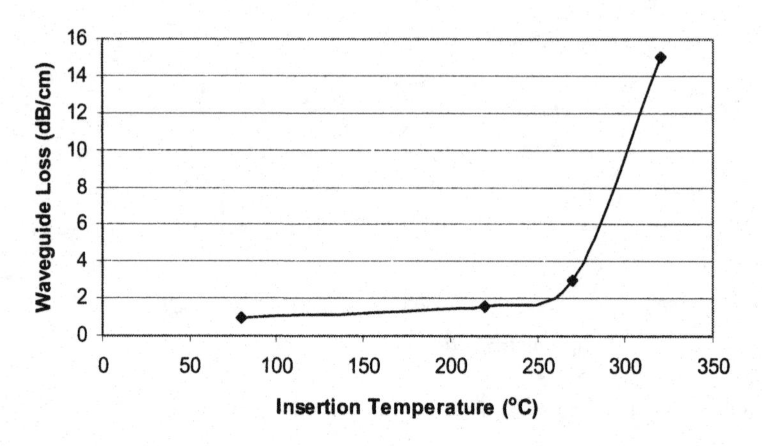

Figure 5: Waveguide Loss as a function of Insertion Temperature

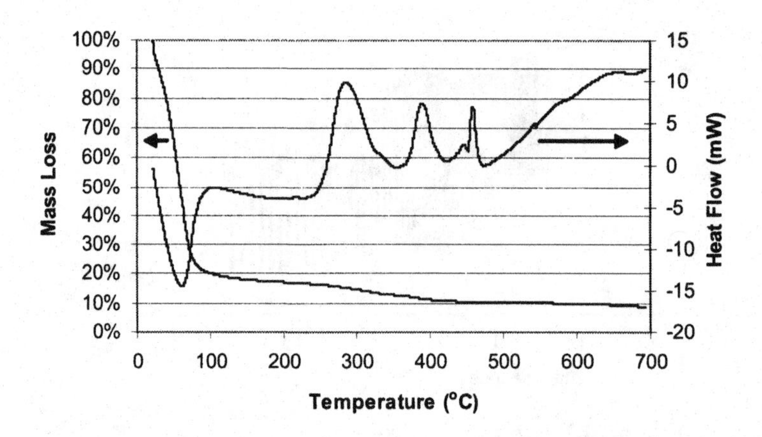

Figure 6: Simultaneous DSC-TGA of Barium Titanium Alkoxide Sol

The index and thickness of the films as function of the insertion temperature is given in figure 7. The samples fired above 300°C evidently begin densifying as demonstrated by the decrease in thickness and increase in refractive index. The sample inserted at 500°C achieved the highest index of refraction, validating the thermal cycle devised by Kozuka, et al[12].

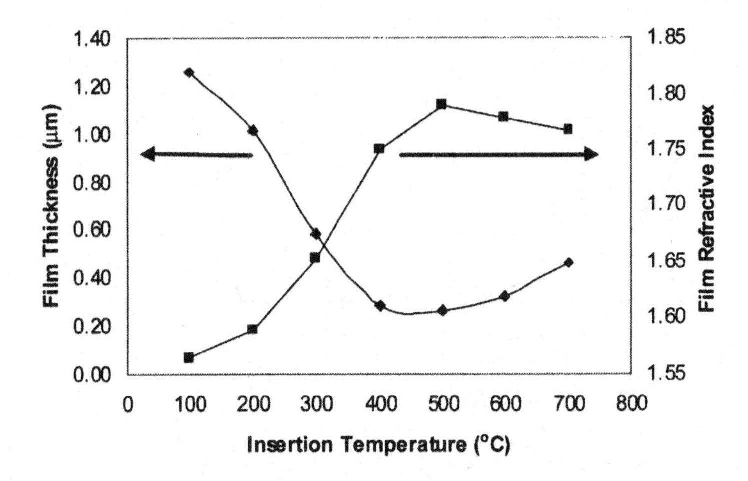

Figure 7: Film Thickness and Refractive Index as a function of Insertion Temperature

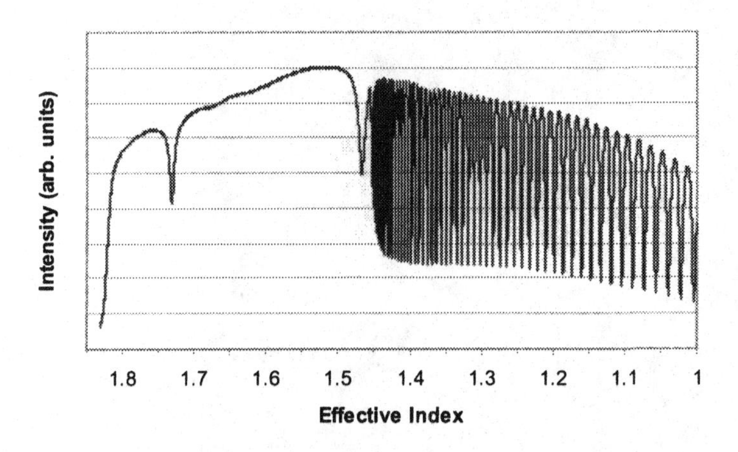

Figure 8: Prism Coupling Spectrum of Fired Sample

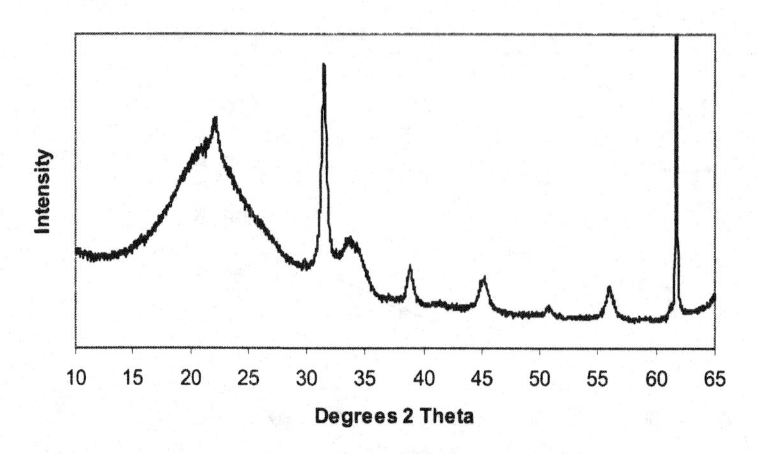

Figure 9: X-ray Diffraction of Fired Sample

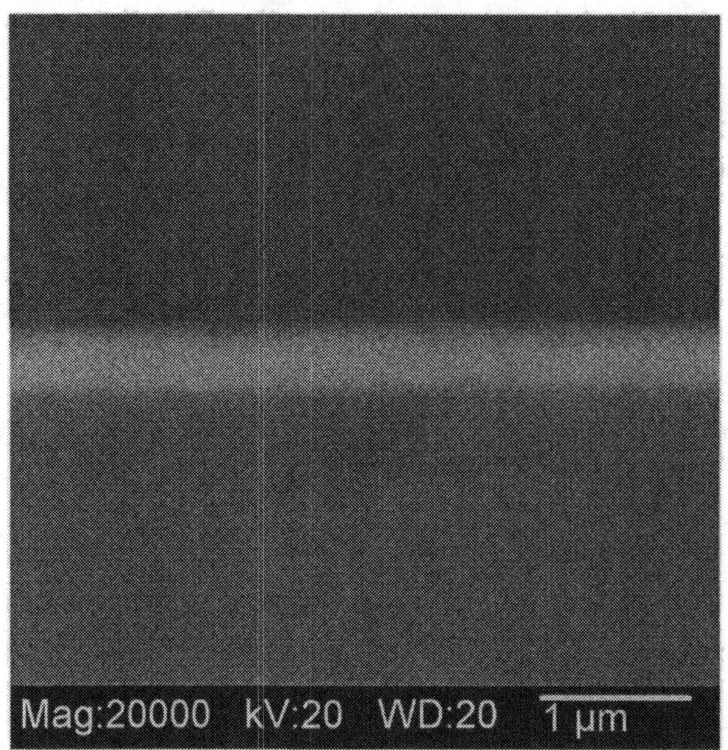

Figure 10: Scanning Electron Micrograph of Fired Sample; Backscattered Mode 20,000x

A sintered sol sample was prepared using optimized processing conditions. The sample was spin cast at 2000 rpm, and inserted at 300°, 500° and 700°C. The prism coupler spectrum of this sample is given in figure 8, indicating the film thickness to be about a third of a micron and an index of 1.90, approaching the bulk value (2.1-2.3). X-ray diffraction was used to probe the crystallinity (figure 9) and crystal size of the film, which was computed to be 10-20 nm. The diffraction data indicates a pseudo-cubic perovskite barium titanate phase with no apparent pyrochlore impurity. Figure 10 is a SEM micrograph taken in backscattered mode. It shows the thickness of the film at approximately half of a micron, consistent with the prism coupling measurement. The roughness of a film was determined to be a low *rms* value of 14 nm, shown in figure 11, using the optical interference method. This is an important factor since scattering of light in the waveguide is influenced by surface roughness[14,15].

Based on the sharp increase in loss at burnout temperatures, it is expected that the transformation of structure of the film is responsible for the loss. Assuming there is little or no absorption in the film as expected, scattering is considered to be the dominant loss mechanism. We propose that upon burnout of the organics and the polymer, the continuous fractal network of the dried gel is lost and nano-sized domains are formed. These domains are defined by their differing connectivities, porosities, residual organic content and hence refractive index, and contribute to scattering. Furthermore, since these domains are expected to be small (nanosized), the drastic increase in loss we report is likely due to scattering by domain clusters whose size is commensurate with wavelength λ. Upon the exposure to higher temperatures, it is believed that these domains template the growth of nano-sized Barium Titanate crystals. The crystal size data collected from diffraction experiments suggests these crystals are in fact 10-20 nanometers in size. The low surface roughness of samples measured also point to this dimension.

The reason we invoke domain/crystal clusters as the loss-causing mechanism is because 20 nm size dielectric moieties are expected to lead to little light scattering by themselves. In other words, the film appears optically microstructured despite have nanograins. Further study of this phenomenon could focus on the response of the films as characterized by any number of standard light scattering techniques[15]. It would also point toward dispersing the grains in a passive media such as high index glass as a means of overcoming this problem. In fact, ferroelectric glass-ceramics are being increasingly investigated currently[16].

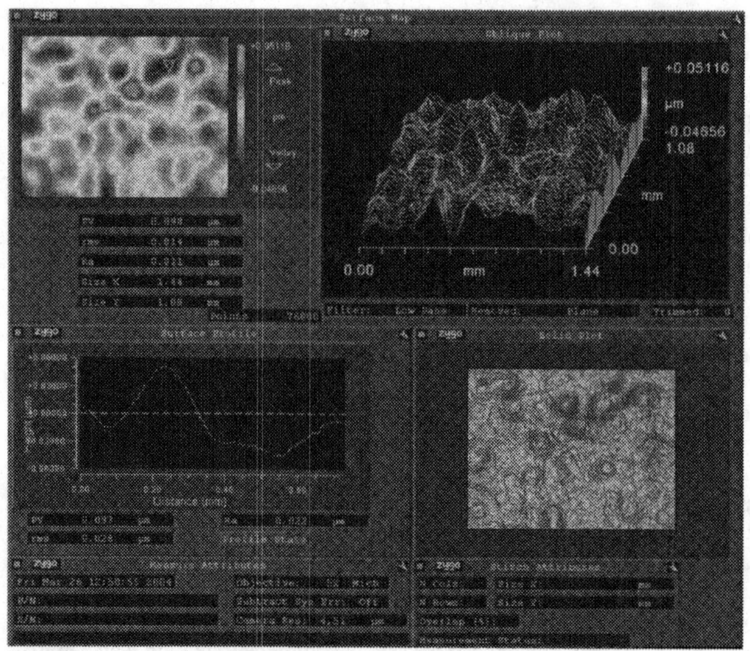

Figure 11: Zygo Interferometer Picture of Fired Sample Surface

CONCLUSIONS

Barium Titanate films were deposited on silica clad silicon wafers. A process was developed to reliably produce crack-free films. X-ray diffraction indicated these films were perovskite barium titanate with no apparent pyrochlore, as desired. Using the X-ray data, we also computed the grains to be in the order of 10-20 nm. The films are also reasonably smooth (rms value of 14 nm) as seen from the measurements using a Zygo interferometer. We were able to control the film thickness by changing deposition parameters with a maximum thickness of a third micron in a single spin. The loss of the gel film was acceptably low (~1 dB/cm). It is when the films were processed that the loss increased. It appears as though the lossy behavior is locked into the structure at low temperatures corresponding to the removal of organics, particularly the polymer polyvinylpyrrolidone from the gel. Further study of the microstructure is warranted to rigorously understand this onset of loss. We propose this low temperature loss in terms of light scattering by clusters formed upon of the burnout of PVP that eventually template the growth of crystals.

REFERENCES

1. D.M. Gill, B.A. Block, C.W. Conrad, B.W. Wessels, and S.T. Ho, "Thin Film Channel Waveguide Electro-optic Modulator in Epitaxial $BaTiO_3$ on MgO," p. 507 in Conference on Lasers and Electro-Optics Europe - Technical Digest, Vol. 11 *Proceedings of the 1997 Conference on Lasers and Electro-Optics.* IEEE, Piscataway, NJ, 1997.
2. M. Siegert, J.G. Lisoni, C.H. Lei, A. Eckau, W. Zander, C.L. Jia, J. Schubert, and C. Buchal, "Epitaxial $BaTiO_3$ Thin Films on Different Substrates for Optical Waveguide Applications," pp. 145-50 in Materials Research Society Symposium - Proceedings, Vol. 597 *Thin Films for Optical Waveguide Devices and Materials for Optical Limiting.* Materials Research Society, Warrendale, PA, 2000.
3. M.B. Sinclair, D. Dimos, B.G. Potter, and R.W. Schwartz, "Light Scattering from Sol-gel Processed Lead Zirconate Titanate Thin Films," *Integrated Ferroelectrics,* **7** [1-4] 225-36 (1995).
4. L. Beckers, W. Zander, J. Schubert, P. Leinenbach, C. Buchal, D. Fluck, and P. Gunter, "Epitaxial $BaTiO_3$ and $KNbO_3$ Thin Films on Various Substrates for Optical Waveguides Applications," pp. 549-54 in Materials Research Society Symposium - Proceedings, Vol. 441 *Thin Films - Structure and Morphology.* Materials Research Society, Pittsburgh, PA, 1997.
5. A. Petraru, M. Siegert, M. Schmid, J. Schubert, and C. Buchal, "Ferroelectric $BaTiO_3$ Thin Film Optical Waveguide Modulators," pp. 279-84 in Materials Research Society Symposium - Proceedings, Vol. 688 *Ferroelectric Thin Films X.* Materials Research Society, 2002.
6. R.A. McKee, F.J. Walker, E.D. Specht, and K.B. Alexander, "MBE Growth and Optical Quality of $BaTiO_3$ and $SrTiO_3$ Thin Films on MgO," pp. 309-14 in Materials Research Society Symposium - Proceedings, Vol. 341 *Epitaxial Oxide Thin Films and Heterostructures.* Materials Research Society, Pittsburgh, PA, 1994.
7. A.R. Teren, S.-S. Kim, S.-T. Ho, and B.W. Wessels, "Erbium-doped Barium Titanate Thin Film Waveguides for Integrated Optical Amplifiers," pp. 413-8 in Materials Research Society Symposium - Proceedings, Vol. 688 *Ferroelectric Thin Films X.* Materials Research Society, 2002.
8. A.M. Glass, "Optical Materials," *Science,* **235** [4792] 1003-9 (1987).
9. H. Kozuka, M. Kajimura, K. Katayama, Y. Isota, and T. Hirano, *in* "Materials Research Society Symposium - Proceedings: Chemical Processing of Dielectrics, Insulators and Electronic Ceramics, Nov 29-Dec 1 1999", Vol. 606, p. 187-92. Materials Research Society, Warrendale, PA, USA, Boston, MA, USA, 2000.
10. H. Kozuka and A. Higuchi, "Stabilization of Poly(vinylpyrrolidone)-containing Alkoxide Solutions for Thick Sol-gel Barium Titanate Films," *J. Am. Ceram. Soc.,* **86** [1] 33-8 (2003).
11. H. Kozuka and A. Higuchi, "Single-layer Submicron-thick $BaTiO_3$ Coatings from Poly(vinylpyrrolidone)-containing Sols: Gel-to-ceramic Film Conversion, Densification, and Dielectric Properties," *J. Mater. Res.,* **16** [11] 3116-23 (2001).

12. H. Kozuka and M. Kajimura, "Single-step Dip Coating of Crack-free $BaTiO_3$ Films >1μm Thick: Effect of Poly(vinylpyrrolidone) on Critical Thickness," *J. Am. Ceram. Soc.,* **83** [5] 1056-62 (2000).
13. D.S. Yoon, J.M. Kim, K.C. Ahn, and K. No, "Effects of Heating Schedule and Atmosphere on the Phase Formation of PLZT Thin Films Prepared Using Sol-Gel Process," *Integrated Ferroelectrics,* **4** 93-101 (1994).
14. B.A. Block, B.W. Wessels, D.M. Gill, C.W. Conrad, and S.T. Ho, *in* "Amorphous and Crystalline Insulating Thin Films Proceedings of the 1996 MRS Fall Meeting,Dec 2-4 1996", Vol. 446, p. 349-54. Materials Research Society,Pittsburgh,PA,USA, Boston,MA,USA, 1997.
15. M.B. Sinclair, D. Dimos, B.G. Potter, R.W. Schwartz, and C.D. Buchheit, "Light Scattering from Sol-Gel $Pb(Zr,Ti)O_3$ Thin Films: Surface versus Volume Scattering," *Integrated Ferroelectrics,* **11** [1-4] 25-34 (1995).
16. M.E. Lines and A.M. Glass, *Principles and Applications of Ferroelectrics and Related Materials*. Oxford University Press, Inc., New York, New York, 1977.

BLUE LIGHT EXCITED GLASSES FOR WHITE LIGHT ILLUMINATION

Yi Zheng and Alexis Clare
New York State College of Ceramics at Alfred University, Alfred, NY 14802

ABSTRACT

Rare earth doped glasses were investigated and their excitation and emission spectra are reported. Energy transfer from Tb^{3+} to Eu^{3+}, and from Dy^{3+} to Tb^{3+} and Eu^{3+} is observed. Through energy transfer, co-doped glasses can be excited under the blue light and emit colors complementary to blue that are needed to produce white light. The increase of blue absorption of host glasses increases the emission of the complementary colors from the glasses. Rare earth doped glasses are potential phosphors for white LEDs.

INTRODUCTION

Lighting is very important to our daily life. In incandescent lighting, lots of energy is dissipated in the air through infrared irradiation. Although fluorescent lamps are much more efficient, their lifetime is short and they use mercury that is toxic. Solid state lighting through light emitting diodes (SSL-LEDs) has the potential to replace traditional incandescent or fluorescent lamps for illumination, because this technique could provide highly efficient, long-lasting, safe and robust lighting devices. There are several ways to produce white light with LEDs.[1] One approach is to mix blue, green and red light from three single-color LEDs. This approach is likely to be the most efficient, as there are no energy conversion losses associated with the phosphors. However, this is also likely to be the most expensive and complex approach, because it has three individual LEDs and each needs its own driving circuit and the LEDs need to be balanced in peak intensity. Another option is a combination of a blue LED and a phosphor. The phosphor emits yellow, or green and red light under the excitation by blue light from the blue LED. The partially transmitted blue light is mixed with the secondary light emitted from phosphor, giving a white light. This approach is less complex. The present commercial white LEDs, with a luminous efficiency 25 lm/W, are based on this approach. They use YAG:Ce^{3+} based inorganic phosphors, which convert the blue light into a very broad yellow emission. However, the luminous efficacy of this type of phosphor is not very high because of its broad emission. Luminous efficacy of optical radiation measures equivalent optical power that can be perceived by human eyes (luminous flux lm/optical power W). It reaches a maximum at about 555 nm in green light region and decreases rapidly for wavelengths approaching both red and blue. Another common problem for inorganic phosphors is the color uniformity. In white LEDs, the phosphor powders are deposited as a film on the blue LED chips. It is very difficult to get a homogeneous dispersion of phosphors in an organic base and to control the thickness of film, which leads to a non-homogeneous color. In contrast, the emission bands due to the f-f transitions of rare earths are narrow, compared to the f-d transition of Ce^{3+}, and the rare earths are homogeneously dispersed in the glass. Therefore, it is possible to design rare earth phosphors which can emit at the wavelength of high luminous efficacy with high homogeneity. In this study, rare earth doped glasses designed for white light illumination under blue light excitation are produced.

EXPERIMENTAL

The host glass A is a sodium borosilicate glass. The glass is chosen by considering the processing conditions under which the glass is coated on the LED chips. The host glasses doped with rare earths Eu_2O_3, Dy_2O_3, Tb_4O_7 and Sm_2O_3 of different concentrations were melted at 1350-1450 °C. The molten glasses were poured into a graphite mold, and annealed at 500 °C for 1 hour. Another host glass B has the similar composition as glass A except for addition of a small amount of phosphate. The results reported here are rare earth doped in glass A except for those indicated specifically. In order to compare the fluorescence intensity, the thickness of the glasses samples was kept constant at 2 mm. The excitation and emission spectra were recorded at front face mode with a Spex Fluorolog 0.22 m fluorometer with a Xenon arc lamp excitation. Transmission spectra were measured using a Perkin Elmer Lambda 900 UV/Vis/NIR spectrometer with blank as reference. The AFM image was take from a Dimension™ 3100 Scanning Probe Microscope (SPM).

RESULT AND DISCUSSION

Single-Doped Glasses

In order to obtain white light using a blue LED excitation approach, the phosphors have to emit complementary colors of blue, i.e., green and red, or yellow light. Several rare earth single-doped glasses were investigated first to determine the color and intensity of their emission in the designated host so that optimized composition of co-doped glasses could be obtained by considering the color balance. It is well known that Eu^{3+} doped glass can emit red light in the visible range.[2] The excitation spectra of Eu^{3+} doped glasses monitored at an emitting wavelength 611 nm are shown in Fig. 1(a). They consist of bands rising from $^7F_0 \rightarrow {}^5D_2$ (~ 464 nm), $^7F_0 \rightarrow {}^5D_1$ (~ 533 nm) and $^7F_0 \rightarrow {}^5D_0$ (~ 579 nm) transitions. The intensity of the excitation bands increases as the Eu_2O_3 concentration increases. There is no concentration quenching for this glass with Eu_2O_3 concentrations up to 3 mol%, which is close to the doping limit in this glass without devitrification. Tb^{3+} doped glasses emit green light efficiently.[3] Fig. 1(b) represents the excitation spectra of Tb^{3+} doped glasses monitored at an emitting wavelength 541 nm. There is an excitation band at about 485 nm in the blue region, however, the excitation wavelength is

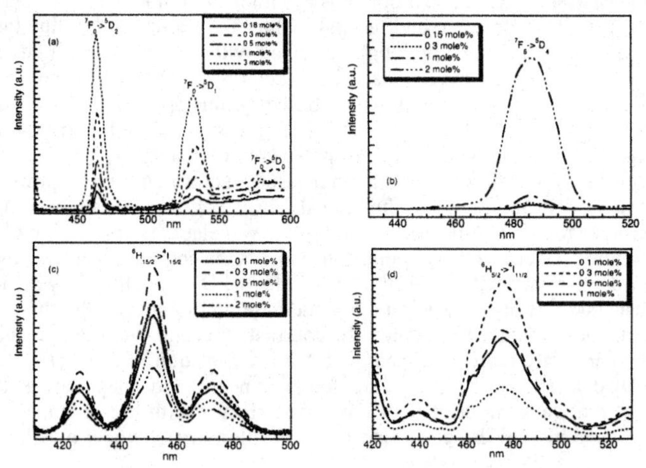

Fig. 1. Excitation spectra of (a) Eu_2O_3(em 611 nm), (b) Tb_4O_7 (em 541nm), (c) Dy_2O_3 (em 572 nm) and (d) Sm_2O_3 (em 596 nm) single-doped glasses

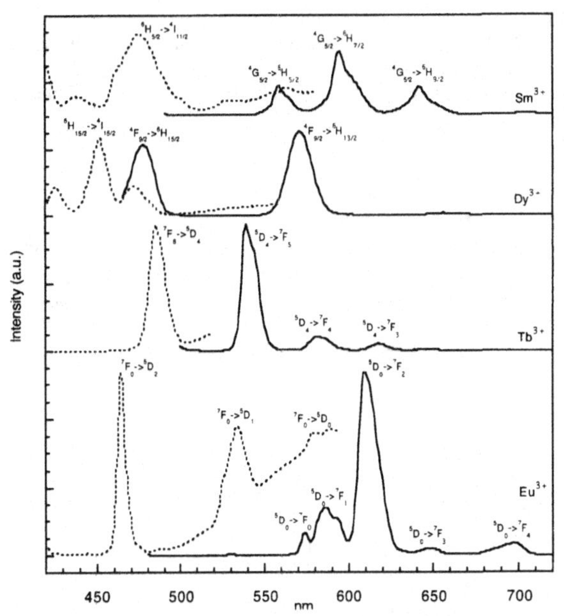

Fig. 2. Excitation (dotted line) and emission (solid line) spectra of 0.3 mol% of Eu_2O_3, Tb_4O_7, Dy_2O_3 and Sm_2O_3 single-doped glasses
Eu^{3+}: em 611 nm, ex 464 nm; Tb^{3+}: em 541 nm, ex 485 nm;
Dy^{3+}: em 572 nm, ex 452 nm; Sm^{3+}: em 596 nm, ex 475 nm

about 20 nm away from major excitation band at 464 nm of Eu^{3+} doped glasses. Tb^{3+} doped glasses show no self quenching when Tb_4O_7 concentration is up added to 2 mol%.

The excitation spectra of Dy^{3+} and Sm^{3+} single-doped glass are given in Fig. 1(c) and (d), respectively. Dy^{3+} doped glasses emit yellow-green light and their excitation spectra are monitored at an emission of 572 nm. The Dy^{3+} doped glasses are excited by irradiation at about 452 nm, which is due to the ${}^6H_{15/2} \rightarrow {}^4I_{15/2}$ transition. The intensity of this excitation band begins to decrease, as Dy_2O_3 concentration is increased above 0.3 mol%. Sm^{3+} doped glasses emit orange light and their excitation spectra are monitored at emission wavelength of 596 nm. Sm^{3+} doped glasses show a broad blue excitation band centered at about 476 nm that is due to the ${}^6H_{5/2} \rightarrow {}^4I_{11/2}$ transition. However, similar to Dy^{3+} doped glasses, the excitation band at 476 nm decreases as the concentration is over 0.3 mol%.

Fig. 2 shows the excitation and emission spectra of 0.3 mol% of Eu_2O_3, Tb_4O_7, Dy_2O_3 and Sm_2O_3 single-doped glasses. The assignments of the observed f-f transitions are given in the figure. From this figure, it is evident that the Eu^{3+}, Tb^{3+}, Dy^{3+} and Sm^{3+} single-doped glasses can be excited by the blue irradiation and emit red, green, yellow-green and orange light separately. Combined with blue light, the emission from a single-doped glass alone cannot yield white color. Therefore, it is necessary to use rare earth co-doped glasses. However, the excitation bands for each Ln^{3+} locate at different wavelengths. The peak wavelength of blue LEDs is 430 to 490 nm, depending on the composition of the LED materials. Typical full width at half maximum of emission is about 30 to 35 nm. In order to maximize the pump efficiency, the peak emission wavelength of blue LEDs has to overlap with the excitation band of rare earth co-doped phosphors as much as possible. This may be achieved by energy transfer between the rare earth ions in co-doped glasses.

Energy Transfer

$Tb^{3+} \rightarrow Eu^{3+}$

Fig. 3(a) shows the excitation spectra of glass co-doped with 0.5 mol% of Eu_2O_3 and 0.2 mol% of Tb_4O_7. When the emission of Eu^{3+} is monitored at a wavelength of 611 nm, an excitation band due to Tb^{3+} at 485 nm appears in the excitation spectrum appearing to attribute to

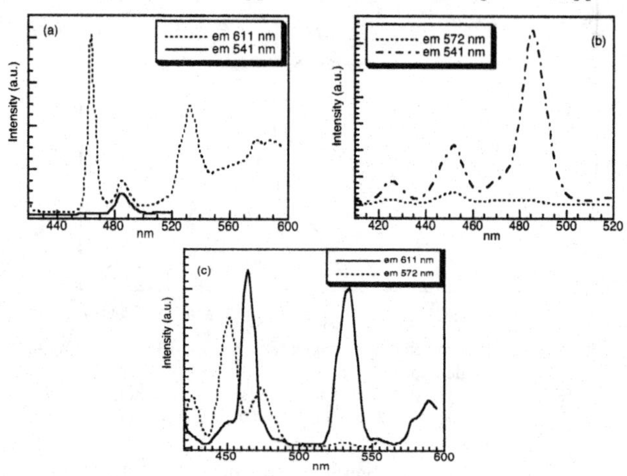

Fig. 3. Excitation spectra of glasses co-doped with (a) 0.5% Eu_2O_3 and 0.2% Tb_4O_7 (b) 0.3% Dy_2O_3 and 1% Tb_4O_7 (c) 0.3% Dy_2O_3 and 1% Eu_2O_3 (mol%)

Eu^{3+} emission. This clearly indicates that there is an energy transfer from the Tb^{3+} ions to Eu^{3+} ions in the glass. According to the Förster-Dexter theory, it is necessary for the absorption band of the acceptor to overlap the emission band of the donor.[4,5] Fig. 2 shows that the Tb^{3+} emission bands $^5D_4 \rightarrow {}^7F_5$ and $^5D_4 \rightarrow {}^7F_4$ overlap Eu^{3+} excitation bands $^7F_0 \rightarrow {}^5D_1$ and $^7F_0 \rightarrow {}^5D_0$, respectively. When the $^5D_4$ level of Tb^{3+} relaxes to $^7F_{5,4}$ levels, the energy transfers to Eu^{3+} ions and they are excited from ground state $^7F_0$ to the $^5D_{1,0}$ levels. Finally, the radiative transitions from $^5D_0$ to the ^7F_J lever occur through non-radiative relaxation from $^5D_1$ to $^5D_0$ level. Through energy transfer, Eu^{3+} and Tb^{3+} co-doped glasses can be excited near 485 nm and emit red and green light simultaneously.

$Dy^{3+} \rightarrow Tb^{3+}$ and Eu^{3+}

The excitation spectra of 0.3 mol% of Dy_2O_3 and 1 mol% of Tb_4O_7 co-doped glass are given in Fig. 3(b). Excitation bands are seen at 426 and 452 nm in the excitation spectrum monitored at Tb^{3+} emission of 541 nm. The position and shape of these two bands are very similar to the excitation bands of Dy^{3+} ions in the spectrum of single-doped glass, so evidently, an energy transfer from Dy^{3+} to Tb^{3+} occurs. The possible energy transfer mechanism is from the $^4F_{9/2} \rightarrow {}^6H_{15/2}$ transition of Dy^{3+} ions to the $^7F_6 \rightarrow {}^5D_4$ transition of Tb^{3+} ions. The energy transfer from

Dy^{3+} to Eu^{3+} is also seen in Dy^{3+} and Eu^{3+} co-doped glass as shown in Fig. 3(c). An excitation band occurs at 452 nm as a shoulder of Eu^{3+} excitation band of 464 nm. This can be predicted from the partial overlap between the Dy^{3+} emission bands ${}^4F_{9/2} \rightarrow {}^6H_{15/2}$ and ${}^4F_{9/2} \rightarrow {}^6H_{13/2}$ and the Eu^{3+} excitation bands ${}^7F_0 \rightarrow {}^5D_2$ and ${}^7F_0 \rightarrow {}^5D_0$, as shown in Fig. 2. Therefore, Dy^{3+}, Tb^{3+} and Eu^{3+} co-doped glasses can be excited by 452 nm.

Although the Dy^{3+} emission band ${}^4F_{9/2} \rightarrow {}^6H_{15/2}$ almost completely overlaps with the Sm^{3+} excitation band ${}^6H_{5/2} \rightarrow {}^4I_{11/2}$, energy transfer is not observed in Dy^{3+} and Sm^{3+} co-doped glasses. The energy involved in these transitions is so close that it could transfer back and forth between these two transitions easily, leading to non-radiative transitions. Another possible reason for no energy transfer is that Dy^{3+} and Sm^{3+} show cross relaxation at very low concentration.

The emission spectrum of Dy^{3+}, Eu^{3+} and Tb^{3+} doped glass is shown as a solid line in Fig. 4(a). The glass can be excited by 452 nm light and emit blue, green, yellow and red light at the same time. The dotted lines in Fig. 4(a) are the simulated blue LED light that would transmit through the glass. This will enable an estimation of the color of the total emission from the glass phosphor. The calculated chromaticities of total emission from this triply co-doped glass are shown as triangles 1 and 2 in the CIE diagram (Fig. 4(c)). The circles in the CIE diagram are the

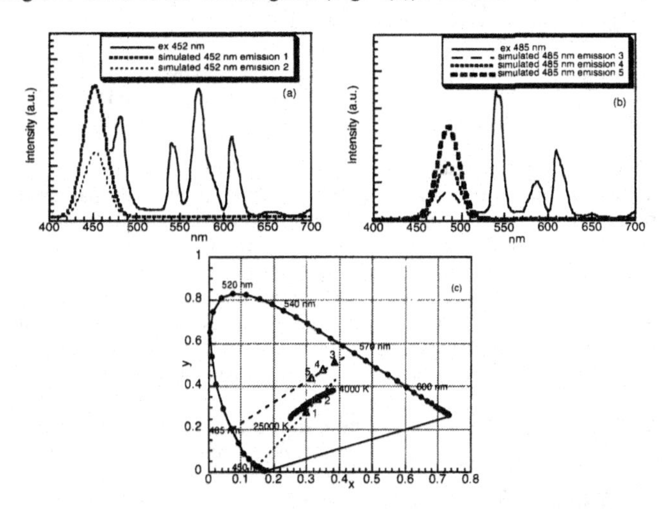

Fig. 4. Emission spectra(the solid lines) and simulated blue transmission light (the dashed lines) of glass co-doped with (a) 0.2% Dy_2O_3, 0.3% Eu_2O_3 and 0.15% Tb_4O_7; (b) 0.5% Eu_2O_3 and 0.2%Tb_4O_7 (mol%); (c) 1931 CIE chromaticity diagram, and chromaticity coordinates of characteristics (Δ) of simulated total emission of glass in (a) and (b)

positions of the day light at different correlated color temperatures from 4000 to 25000 K.[6] These circles form a black body curve. Any points on or near the curve of the day light represent white color. The locations of the simulated total emission of Dy^{3+}, Eu^{3+} and Tb^{3+} doped glass are very close to the black body curve of the day light. The solid line in the Fig. 4(b) is the emission spectrum of Eu^{3+} and Tb^{3+} co-doped glass. The glass emits green and red colors simultaneously under excitation of the blue light at 485 nm. The dotted lines are the simulations of the transmitted blue light in different intensity. The calculated chromaticities of total emission are shown as triangles 3, 4 and 5 in the CIE diagram. Since red emission is not intense enough, they are far away from the black body curve of the day light, indicating the glass with this composition can not produce white light under blue excitation at 485 nm. The composition has to be adjusted so that it can emit more red light, and then the total emission can be white color.

Effect of Host Glasses

For glass phosphors to produce white light under blue excitation, most of the blue light has to be absorbed and down-converted to red and green, or yellow light by the phosphors, so they need have high concentration of high absorption-coefficient center in the blue range. If not, then the phosphors have to be very thick to absorb enough blue light. In general, most materials absorb

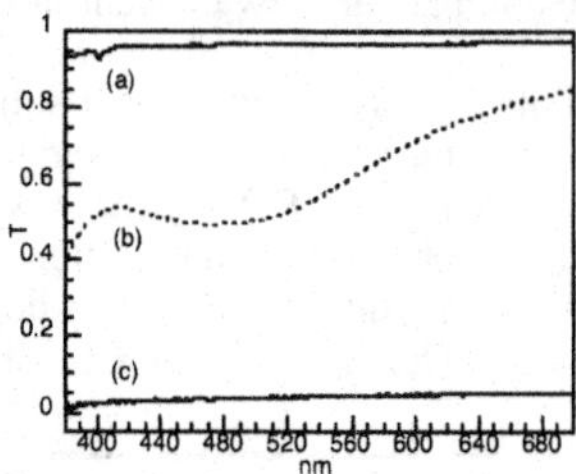

Fig. 5. Transmission spectra of (a) glass A, 0.18% Eu_2O_3, (b)glass A, 0.18% Eu_2O_3, 0.56% MnO, (c) glass B, 0.73% Eu_2O_3

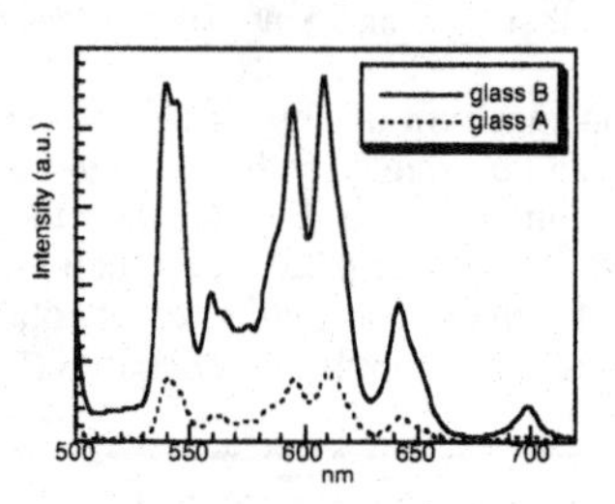

Fig. 6. Emission spectra of 0.3% Eu_2O_3, 0.3% Sm_2O_3, 0.2% Dy_2O_3 and 0.15% Tb_4O_7 (mol%) co-doped glass A and B, ex 483 nm

more light as the wavelength extends further into the UV range. Therefore, the absorption of exciting light is usually not a problem for fluorescent lamp phosphors that are excited by UV light, but the rare earth ions absorb blue light poorly. Although complementary colors of blue can be obtained under the blue irradiation, the low absorption at blue region is a barrier for high luminous efficiency and yielding a white light by color balance. Fig. 5 shows the transmission spectra of Eu^{3+}, Mn^{2+} doped glasses A and B. Eu^{3+} doped glass A absorbs only lightly in the blue region. Although Eu^{3+} and Mn^{2+} co-doping increases the absorption in the blue region, the intensity of Eu^{3+} emission decreases significantly (the emission spectrum not shown). The transmission of Eu^{3+} doped glass B is very low across the visible regions.

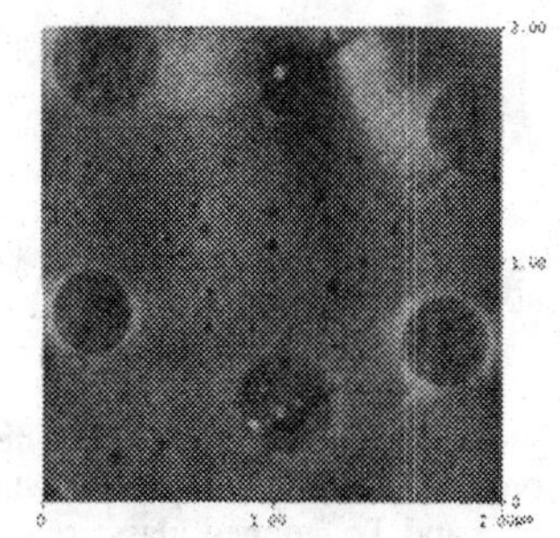

Fig. 7. AFM image of glass B co-doped with 0.3% Eu_2O_3, 0.3%, Sm_2O_3, 0.2% Dy_2O_3 and 0.15% Tb_4O_7 (mol%)

The emission intensity from phosphors is determined by their quantum efficiency and absorption of exciting light. The latter is proportional to the absorption coefficient and the optical path length. The higher the absorption and the quantum efficiency, the stronger the emission. In order to increase the emission from phosphors, the absorption in the blue range must increase. Fig. 6 compares the emission spectra of rare earth co-doping in the two host glasses A and B. The emission intensity increases nearly five-fold in glass B, which implies an increase in the absorption of the exciting blue light and/or quantum efficiency. Host glass A is a transparent sodium borosilicate glass, which absorbs only a very small portion of the blue light. Glass B is a translucent sodium boro-phospho-silicate glass. From the AFM image (Fig. 7), glass B is liquid-liquid phase separated with a droplet structure in the glass matrix. The average diameter of droplet is close to the wavelength of blue light. The incident light is scattered by the droplets and this increases the optical path length, giving more opportunity for the absorption of the incident light in glass B. Since the size of scattering center is much smaller than wavelength of the green, yellow and red light, the emitted light is not scattered as much as the blue light. Another possible reason for increase of emission intensity in glass B is that emission may be more efficient in this glass as a result of the added phosphate.

CONCLUSION

The excitation and emission spectra of rare earth single- and co-doped glasses are reported. Through energy transfer from Tb^{3+} to Eu^{3+}, Tb^{3+} and Eu^{3+} doped glass can emit green and yellow color under excitation near 485 nm. Dy^{3+}, Tb^{3+} and Eu^{3+} co-doped glasses can be excited by 452 nm irradiation and emit multi colors through energy transfer from Dy^{3+} to Tb^{3+} and Eu^{3+}. High absorption of excitation source of the glasses has the effect on the increase of emission intensity.

We acknowledge the financial support of New York State Office of Science, Technology and Academic Research for this study.

REFERENCE

[1]"Light Emitting Diodes (LEDs) for General Illumination - An OIDA Technology Roadmap Update 2002", edited by J. Y. Tsao, published by Optoelectronics Industry Development Association.

[2]E.W.J.L. OOMEN and A.M.A. van DONGEN, "Europium (III) in Oxide Glasses, Dependence of the Emission Spectrum upon Glass Composition", Journal of Non-Crystalline Solids, **111** (1989) 205-213,

[3]G. Amaranath and S. Buddhudu, "Spectroscopic Properties of Tb^{3+} Doped Fluoride Glasses", Journal of Non-Crystalline Solids **122** (1990) 66-73

[4]D. L Dexter, J. Chem. Phys. **21** (1953) 836

[5]Th. Förster, Ann. Physik **2** (1948) 55

[6]CIE (Commission International d'Eclairage) data 1931

THE DEVELOPMENT OF AN ARSENIC SULFIDE GLASS BASED PHOTORESIST

Lauren Russo, Miroslav Vlcek* and Himanshu Jain
Department of Materials Science and Engineering
Lehigh University
Bethlehem PA 18105

Madan Dubey
Army Research Laboratory
Adelphi, MD 20783

ABSTRACT

The photoinduced changes in the structure of chalcogenide glasses produce a selective etching effect, which has been investigated for the development of a photoresist. The phenomenon of photoinduced silver diffusion into the chalcogenide film was also used to make the photoresist more resistant to a chemical etchant so that selectivity was greatly enhanced. The performance of binary and silver doped chalcogenide photoresists was evaluated and compared by examining their in-situ etching kinetics and resolution capability.

INTRODUCTION

Photoresists play an important role in the fabrication of three-dimensional microscopic structures in microelectronics and microelectromechanical systems (MEMS). There are several important characteristics that are necessary for any material to be used as a photoresist. In order to achieve the required fine resolution of the features, it is important that the photoresist material be amorphous with a uniform structure.[1] Another important characteristic for the performance of a photoresist is its selectivity, which is the ratio of the etching of the unexposed and exposed regions of the film.[2] In order to produce a good photoresist mask, one needs high selectivity with a large difference between the etching rates for the regions that were exposed to, and that were not exposed to light. If the illuminated area of the photoresist dissolves at a slower rate than the unexposed areas, one has a negative photoresist. Research has shown that the opposite effect, positive etching, can also be achieved in the same chalcogenide films by altering the solvent used in dissolution.[2,3] All the results presented in this paper deal with the evaluation of a negative photoresist. The selectivity and overall performance of a photoresist are evaluated in terms of the kinetics of the photoinduced changes as well as the kinetics of the film dissolution.

Currently the most popular materials for lithography are organic polymer photoresists. However, due the increasing reduction in size and geometry, present research is focused on the development of chalcogenide glass photoresists.[1,4] Most chalcogenide glasses are photosensitive materials. Exposure to light causes structural changes that allow their properties to be easily manipulated. A very important photoinduced effect that is pertinent to the material's use as a photoresist is the selective etching that is produced by the change in chemical solubility when exposed to light. When chalcogenide glass film is deposited onto a substrate through thermal evaporation, a variety of bonding and coordination defects are produced.[5-7] However, through exposure to light, the local structure is altered such that the defects recombine and the various atoms approach proper coordination and configuration.[8] This modification causes the chalcogenide layer to have a higher or lower activation energy for dissolution in a solvent, which eventually becomes the source of selective etching.[9]

There are potentially several benefits of using chalcogenide glass as a photoresist in microlithography. Typically, thin films of chalcogenide glass are applied to a substrate using

* On leave from University of Pardubice, 532 10 Pardubice, Czech Republic

thermal vacuum evaporation, which minimizes any contamination of the film or substrate. The use of an orbital rotating deposition stage produces a uniform thickness. One of the most significant advantages is that thinner layers can be applied to protect the underlying regions of the substrate because the chalcogenide film is highly resistant to acids that are used to create 3D structures in the substrates. With thinner layers the problems associated with the interference of the incoming light within the photoresist film are eliminated. Therefore, the illumination from the light source is more precise and one obtains sharper resolution at the pattern edges with the elimination of interference effects.[4,10]

We have also focused on increasing the chemical resistance of a chalcogenide photoresist through the use of photoinduced diffusion of silver into the film. The silver deposited as a layer on top of the chalcogenide film can be diffused into the photoresist via direct exposure to light or flood exposure through some type of mask. This diffusion of silver causes the formation of a new type of silver containing glass in the exposed regions. It has been proven generally that silver-chalcogenide glasses of these compositions are chemically more resistant to alkaline solution than the starting chalcogenide composition.[11,12]

In short, the goal of this research is to evaluate the performance of As-S glass photoresists, including the in-situ etching kinetics and the resolution capability. Comparisons are then made regarding the effect of processing variables as well as the performance of an As-S photoresist versus that of a silver photodiffused composition.

EXPERIMENTAL PROCEDURE

Material Preparation

The base material used in this study was $As_{35}S_{65}$ as a thin film on a glass slide substrate. The films were prepared by thermal vacuum evaporation at 1.3×10^{-6} Pa. For the silver photodoped samples, a thin film of silver was deposited on top of the As-S film by thermal vacuum evaporation.

Silver Doped Chalcogenide Photoresist

There are several steps involved in creating a patterned photoresist to protect the selected regions of the substrate during etching. The steps used to develop a photoresist, which takes advantage of the photoinduced diffusion of silver, are illustrated in Figure 1.

Exposure to a halogen white light through the openings in a patterned mask caused the photoinduced diffusion of silver into the selected regions of the chalcogenide layer, creating a new silver containing glass composition. The remaining metallic silver, which was not diffused into the chalcogenide film was then removed by washing in a nitric acid solution. The photoresist was then developed through submersion in a non-aqueous, amine based, chemical solvent. The result was a photo-doped silver-chalcogenide photoresist mask to be used for further etching into the substrate to create three-dimensional structures.

Etching Kinetics

The first important parameter in evaluating the performance of a photoresist is the kinetics of dissolution while etching. From this information one can determine the photoinduced resistance of a material to a chemical etchant, the etching rate of the material, as well as the potential selectivity. The etchant was a non-aqueous, amine based solution. The changes in the thickness of the film during etching were monitored through the use of a monochromatic light source at 632.8 nm, with the sample suspended in the etching bath as seen in Figure 2. A detector then recorded the changes in transmitted light intensity over time. As the thickness of the film decreased, the transmitted light intensity oscillated because of the interference of the incoming light until the film was completely dissolved. This experimental setup was used to compare the photoinduced chemical resistance of the samples of $As_{35}S_{65}$ after having been exposed to a halogen light source (14 mW/cm^2) for different exposure dosages. The chemical resistance of the binary chalcogenide films was compared to that of a silver-chalcogenide film created through the photoinduced diffusion of silver.

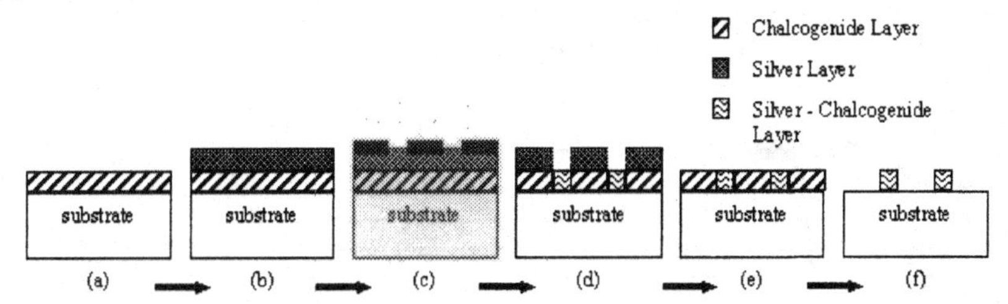

Figure 1: Schematic illustration of the development of a silver doped chalcogenide photoresist

Etching Resolution

For evaluating the resolution capability of the patterns that can be produced in a photoresist, we exposed our films to halogen white light through a chromium mask containing features of varying size and separation. Light optical microscopy was used to analyze the resulting structures in the film. The regions where light passed through the mask into the chalcogenide layer became more resistant to the etchant when the photoresist was developed. The two variables that were varied in this process were the thickness of the photoresist layer (1 μm and and 0.33 μm) and the exposure time that ranged between 0 and 10 minutes.

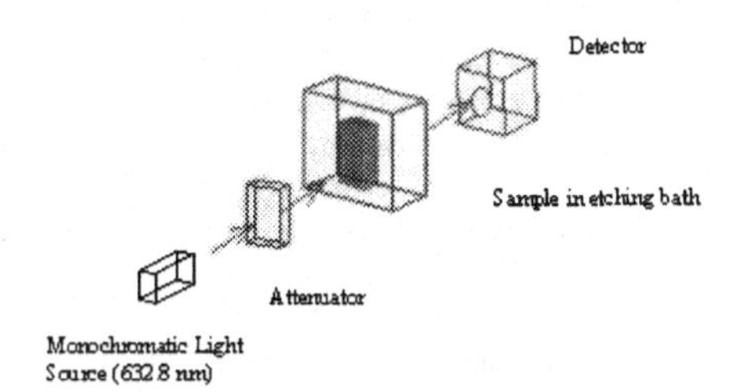

Figure 2: Schematic diagram of the setup for the in-situ measurement of etching kinetics.

RESULTS AND DISCUSSION

Etching Kinetics

The periodic variation of transmitted intensity caused by the decreasing thickness of the sample in the chemical etchant can be seen as a function of time in Figure 3. We determined the time for the sample to completely dissolve in the etchant from the point where the transmission stopped oscillating. It is concluded that the samples that had higher illumination dosages were more resistant to the chemical etchant because they took longer to fully dissolve. This data can be further analyzed by using the peaks in transmitted light intensity to plot the change in thickness with time, using the following equation:

$$2n\Delta d = \Delta k\lambda \quad (1)$$

where λ is the wavelength of the monochromatic probe light (λ = 632.8 nm) and n is the refractive index of the $As_{35}S_{65}$ film estimated to be 2.4. Then from Equation 1, the change in thickness (Δd) can be calculated for the time span between maximum peak and minimum of the valley (Δk=0.5) in Figure 3. From the plot of change in thickness versus time, it is easy to visualize the kinetics of etching as well as potential selectivity (see Figure 4).

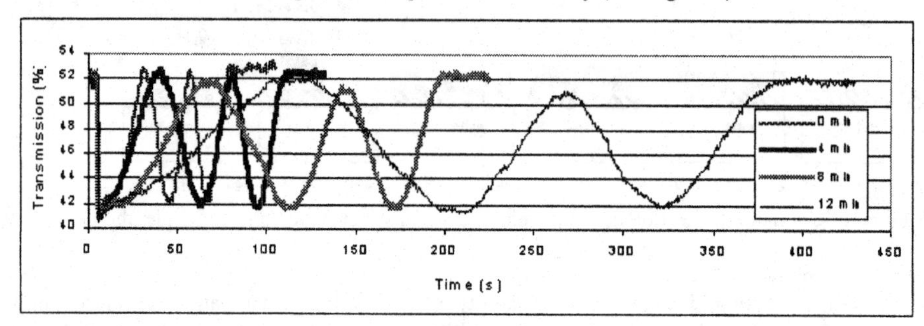

Fig. 3 Time dependence of transmission changes at λ = 632.8 nm during etching of different time exposed $As_{35}S_{65}$ thin layers.

From Figure 4, several observations can be made regarding the etching rates and selectivity exhibited by the photoresist as exposure dosage is varied. The etching rate is represented by the slope of the corresponding lines. The data show that with increased exposure to light, the average etching rate of the photoresist decreases. This decrease is due to the photoinduced changes in the chalcogenide film, which enhance its chemical resistance to the etchant. The most important feature to note in this figure is the line that represents the etching of the silver photodoped sample. There is no change in its thickness because it was fully resistant to the chemical solvent. Note that it only required a 30 second exposure to light for the silver to diffuse significantly enough into the chalcogenide film to make it completely resistant to the etchant.

In order to achieve maximum selectivity, the highest possible difference in etching rates is needed between the exposed and unexposed regions of the film. We find that the difference in the two etching rates and hence selectivity increases with increasing illumination dosage. Practically complete selectivity is obtained for the silver photodoped sample. In this case the regions that underwent photoinduced diffusion of silver are fully protected from the chemical etchant while the unexposed regions dissolve away quickly.

Etching Resolution

We found that film can be etched to obtain photoresist mask with micrometer resolution through optimal control of variables. We determined the effect of film thickness and exposure dosage on etching resolution. Figure 5 compares the features in the photoresist produced by varying these parameters. It shows that a small variation in the

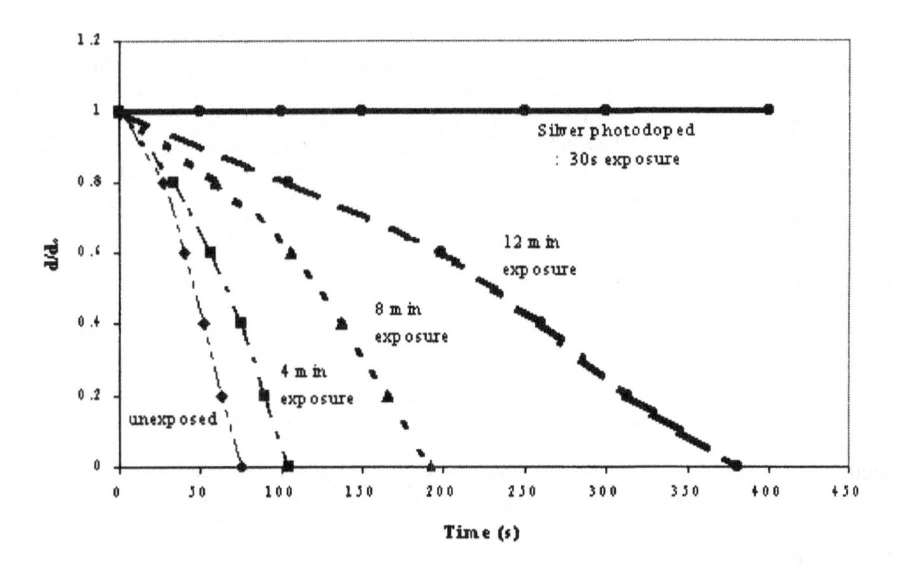

Figure 4: The etching kinetics of $As_{35}S_{65}$ samples exposed for varying times (d_0 = 330 nm).

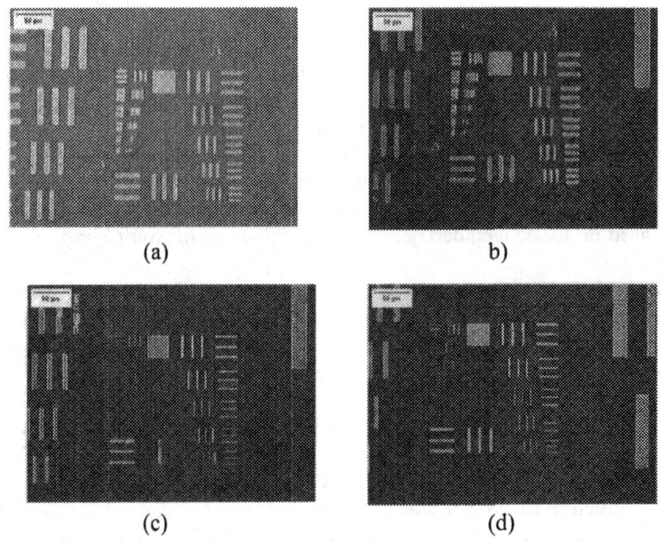

(a) b)

(c) (d)

Figure 5: Patterns formed in $As_{35}S_{65}$ photoresist from exposure through a mask. Starting thickness: 1 μm (a, c) and 330 nm (b,d). Exposure dose 10 min (a, b) and 6 min (c, d)

parameters can cause significant deviation in the photoresist performance. Figures 5a and 5b illustrate that if the exposure dosage is too high, there is significant intensity of diffracted light, which is detrimental especially in the regions of the smallest slits of the mask. Because of this problem, the sharp resolution at the edges is not achievable. On the other hand, if the illumination time is too short, the sharp resolution is obtained but then the exposure dosage is not high enough to protect the full depth of the film. So the smaller features are washed away in the etchant, as was the case for the photoresist shown in Figure 5c. Therefore, for the optimal performance of a photoresist, it is important to balance the effects of the film thickness with the exposure dosage. An example of successful balance is seen in Figure 5d, where the photoresist has all of patterns as well as sharp resolution.

CONCLUSION

From the present work, we find that chalcogenide glasses have several desirable qualities for the photoresist applications in microlithography. They offer all the desirable characteristics of traditional organic polymer photoresists, with potential for further improvements. For example, through the use of the photoinduced diffusion of silver, the photoresist exhibits superior selectivity. In addition, chalcogenide photoresists yield high etching resolution for making patterns that are necessary for fabricating three-dimensional structures into substrates.

ACKNOWLEDGEMENT

Separate parts of this work were supported by the National Science Foundation (DMR 00-74624 and DMR 03-12081) and a Lehigh-Army Research Lab (ARL) collaborative research project.

REFERENCES

[1] M. Vlcek, M. Frumar, M. Kubovy, V. Nevsimalova, "The Influence of the Composition of the Layers and of the Inorganic Solvents on Photoinduced Dissolution of As-S Amorphous Thin Films," *Journal of Non-Crystalline Solids*, **137-138** 1035-1038 (1991).

[2] M. Vlcek, J. Prokop, M. Frumar, "Positive and Negative Etching of As-S Thin Layers," *International Journal of Electronics*, **77** 969-973 (1994).

[3] N.P. Eisenberg, M. Manevich, and M. Klebanov, "Fabrication and Testing of Microlens Arrays for the IR Based on Chalcogenide Glassy Resists," *Journal of Non-Crystalline Solids*, **198-200** 766-768 (1996).

[4] M. Nakase, and Y. Utsugi, "Exposure Characteristics and Proximity Effect in $Ag_2Se/GeSe_4$ Inorganic Photoresists," *Journal of Vacuum Science and Technology*, **A3** 1849-1854 (1985).

[5] J. Li, D.A. Drabold, S. Krishnaswami, G. Chen, and H. Jain, "Electronic Structure of Glassy Chalcogenides As_4Se_4 and As_2Se_3: A Joint Theoretical and Experimental Study," *Physics Review Letters*, **88** 046803-1 (2002).

[6] H. Jain , "Comparison of Photoinduced Atom Displacement in Glasses and Polymers," *Journal of Optoelectronics and Advanced Materials*, **5** 5-22 (2003).

[7] K. Antoine, J. Li, D. A. Drabold, H. Jain, M. Vlcek, and A.C. Miller, "Photoinduced changes in the electronic structure of As_2Se_3 glass," *Journal of Non-Crystalline Solids*, **326-327** 248-256 (2003).

[8] G. Chen, H. Jain, S. Khalid, J. Li, D. A. Drabold, and S. R. Elliott, "Study of Structural Changes in Amorphous As_2Se_3 by EXAFS under In-Situ Laser irradiation," *Solid State Communication*, **120** 149-153 (2001).

[9] K. Petkov, M. Vlcek, and M. Frumar, "Photo- and Thermo-Induced Changes in the Properties of Thin Amorphous As-S Films," *Journal of Materials Science*, **27** 3281-3285 (1992).
[10] H. Nagai, A. Yoshikawa, Y. Toyoshima, O. Ochi, and Y. Mizushima, "New Application of Se-Ge Glasses to Silicon Microfabrication Technology," *Applied Physics Letters*, **28** 145-147 (1976).
[11] M. Marcus, A. Wagner, and T. Hanley, "Optical Monitoring of Development Kinetics of $GeSe_2$," *Proceedings of the Symposium on Inorganic Resist Systems*, **9** 295-302. (1982).
[12] A. Yoshikawa, O. Ochi, H. Nagai, and Y. Mizushima, "A Novel Inorganic Photoresist Utilizing Silver Photodoping in Selenium-Germanium Glass Films," *Applied Physics Letters*, **29** 677-669 (1976).

STUDY OF RESBOND® CERAMIC BINDERS USED FOR HIGH TEMPERATURE NON-CONTACT THERMOMETRY

S.M. Goedeke*, W.A. Hollerman**, N.P. Bergeron**, S.W. Allison*, M.R. Cates*, T.J. Bencic†, C.R. Mercer†, and J.I. Eldridge§

* Engineering Science and Technology Division, Oak Ridge National Laboratory (ORNL), P.O. Box 2008, M.S. 6054, Oak Ridge, Tennessee 37831

** Department of Physics, University of Louisiana at Lafayette, ULL Box 44210, Lafayette, Louisiana 70504

† Optical Instrumentation Technology Branch, NASA John H. Glenn Research Center at Lewis Field, 21000 Brookpark Road, Mailstop 77-1, Cleveland, Ohio 44135-3191

§ Environmental Durability Branch, NASA John H. Glenn Research Center at Lewis Field, 21000 Brookpark Road, Cleveland, Ohio 44135

ABSTRACT

Fluorescence is a good non-contact thermometry technique in hostile environments such as those found at high temperatures. Phosphors are typically rare earth-doped ceramics that emit light when excited. The intensity, rise time, decay time, and wavelength shift of this emitted light can be temperature dependent. When thermographic phosphors are applied to a surface, with an excitation source and a method to characterize the emission provided; it is possible to determine the surface temperature. One of the simpler methods to apply these coating is through the use of temperature sensitive paints (TSPs). These TSPs are created by mixing phosphor with a binder material to form a sprayable coating that can be easily and economically applied to a large area. Ideally, these phosphor paints need to survive at the limit of the existing decay time data, or 1700 °C. The survivability of phosphor paint depends on the physical characteristics of the binder. The goal of this research is to discover binders that will allow phosphor paints to survive at high temperatures. Suitable binders will allow the construction of non-contact measurement devices useful in environments that are not suited for more common thermocouple or infrared devices. For a phosphor paint to be useful at a selected temperature, it must fluoresce when excited and have a measurable decay time. In this study, the Cotronics Resbond® 791, 792, and 793 ceramic binders were evaluated to determine their suitability to serve as binders for a Y_2O_3:Eu phosphor powder. Post-thermal cycling spectral analysis was used to quantify wavelength and intensity changes in emission from ultraviolet excitation. Several of the paints utilizing their binders were able to survive at 1500 °C.

INTRODUCTION

Phosphors are fine powders that are doped with trace elements that give off visible light when suitably excited. Many of these phosphors have a ceramic base and can survive and function at high temperatures such as those present during combustion. When thee phosphor is applied as a thin coating, it quickly equilibrates to the ambient environment and can be used to measure the surface temperature.

The basic physics of thermographic phosphors is well established, and researchers at Oak Ridge National Laboratory (ORNL) have demonstrated several useful applications [ref. 1-12]. The thermometry method relies on measuring the rate of decay of the fluorescence yield as a function of temperature. Having calibrated the phosphor over the desired temperature range, a small surface deposit is excited with a pulsed laser. The resulting fluorescent decay (typically in less than 1 ms) time is measured to calculate the temperature of the substrate. In many instances, (e.g., in a continuous steel galvanneal process) a simple puff of powder onto the surface provides an adequate fluorescent signal [ref. 1-2].

Often temperature measurements are made using thermocouples or optical pyrometry. However, in situations where rapid motion or reciprocating equipment is present at high temperatures, it is best to use other techniques. For many phosphors, the prompt fluorescence decay time (τ) varies as a function of temperature and is defined by:

$$I = I_0 \exp\left\{-\frac{t}{\tau}\right\}, \tag{1}$$

where:

I	=	Fluorescence light intensity (arbitrary units),
I_0	=	Initial fluorescence light intensity (arbitrary units),
t	=	Time since cessation of excitation source (s), and
τ	=	Prompt fluorescence decay time (s) [ref. 1-2].

The time needed to reduce the light intensity to e^{-1} (36.8%) of its original value is defined as the prompt fluorescence decay time. An example of this quantity for several thermographic phosphors is shown in Figure 1. Notice the fluorescence decay time decreases by four orders of magnitude when the temperature increases from 600 to 1100 °C.

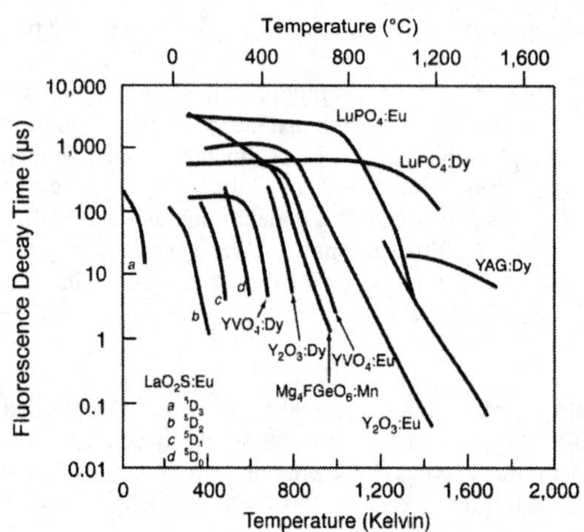

Figure 1. Prompt fluorescence decay time for a selection of thermographic phosphors.

This paper will give an overview into research to find binder and phosphor combinations that can emit fluorescence and remain mechanically viable at temperatures as large as 1600 °C. Emphasis will be placed on developing procedures and techniques for the application and the pre-treatment of candidate binder and phosphor combinations.

HISTORY

In May 2001, a research program was initiated to evaluate TSPs that could be used in high temperature thermometry applications. This three-year program was funded by the NASA John H. Glenn Research Center in Cleveland, Ohio. A research team lead by ORNL was assembled to locate binder and phosphor combinations that can emit fluorescence and maintain structural integrity at high temperatures. The ultimate goal for this research was to locate TSPs that could be used to measure the heat flux and temperature present inside a common turbine or rocket engine. The phosphors in question are typically rare earth compounds that emit copious fluorescence and have grain sizes of less than 10 μm. Binders used for this purpose must: 1) Be easy to apply with an airbrush, 2) Set to a temperature resistant inorganic finish, and 3) Have minimal reaction to the phosphor material.

During the first year of the program, research was completed to determine if heat flux could be measured using a phosphor coating. In the second year, research was completed to find phosphor and binder combinations that can emit light and remain mechanically viable at high temperatures. Results from these measurements can be found in Table I. Each paint combination was exposed to the high temperature listed in Table I for one hour and then slowly cooled. The heating rate was kept small in order to minimize effects due to the difference in expansion coefficient between the paint and the ceramic substrate.

Table I. Zyp Coatings-based thermally sensitive paint survivability results.

Binder	Paint		Emission (nm)	Phosphor Emission At Given Cycling Temperature?					Comments
	Fraction (Vol. %)	Phosphor		1200 °C	1300 °C	1400 °C	1500 °C	1600 °C	
HPC	20%	Y_2O_3:Eu	611			Yes	Yes		Paint mostly gone after heating to 1500 °C
HPC	10%	Y_2O_3:Eu	611			Yes	Yes		Paint mostly gone after heating to 1500 °C
75% HPC 25% LK	20%	Y_2O_3:Eu	611				No		All of paint gone after heating to 1500 °C.
75% HPC 25% LK	10%	Y_2O_3:Eu	611				Yes		Most of coating gone after heating to 1500 °C.
50% HPC 50% LK	20%	Y_2O_3:Eu	611			Yes	Yes		Paint mostly gone after heating to 1500 °C
50% HPC 50% LK	10%	Y_2O_3:Eu	611			Yes	No		No 611 nm peak after heating to 1500 °C.
ZAP	50%	Y_2O_3:Eu	611	Yes	Yes	Yes	Yes		Paint still intact with diminished fluorescence after heating to 1500 °C.
ZAP	30%	YAG:Dy	585	Yes	Yes	Yes	Yes	Yes	Fluorescence still present after heating to 1600 °C.
ZAP	30%	YAG:Tm	420 480				Yes		Coating intact but poor fluorescence after heating to 1500 °C.
ZAP	30%	YAG:Eu	595 611				Yes		Coating intact but poor fluorescence after heating to 1500 °C.

Three binders, HPC®, LK®, and ZAP®, from Zyp Coatings, Incorporated of Oak Ridge, Tennessee were used in this research [ref. 9-10]. Each of these three binders was designed for use in high temperature applications and are available in the commercial market. A selection of rare earth compounds are used for the phosphor powders in this research: yttrium oxide doped with europium (Y_2O_3:Eu), and YAG doped with dysprosium (YAG:Dy), terbium (YAG:Tb), and europium (YAG:Eu). A selected binder and phosphor powder are mixed together to create a strong and durable thermally sensitive paint (TSP). Detailed information on this research can be found in references 9 and 10.

Results shown in Table I show that most of the HP/LK bound samples survived heating to 1400 °C. Most of the paints were removed from the surface of the ceramic at 1500 °C. This data also shows that paint consisting of a 100% ZAP binder and 30% YAG:Dy powder by volume is intact and emits fluorescence after heating to 1500°C. This paint surface did look bumpy or mottled after heating. The other ZAP paints were also intact, but with reduced fluorescence. In fact, the ZAP binder and 30% by volume YAG:Dy TSP was also found to emit fluorescence after heating to 1600 °C.

BINDERS

During the third and final year of the NASA-Glenn research, several Resbond® ceramic binders manufactured by Cotronics Corporation of Brooklyn, New York were evaluated using a powdered Y_2O_3:Eu phosphor. The main fluorescence emission from Y_2O_3:Eu occurs at a wavelength of 611 nm. Selected material and preparation properties for the tested Resbond binders are given in Table II. The Cotronics series of ceramic binders includes six different formulations. Of these, only the four varieties listed in Table II was tested. During initial research, Cotronics Resbond 795 did not adhere to the surface. Since the other two binders have similar properties, it was considered more productive to focus on the three remaining binders.

Table II. Properties for selected Cotronics Resbond® ceramic binders.

Resbond Composition	**791 Silicate Glass**	**792 Silicate Glass**	**793 Silica Oxide**	**795 Alumina Oxide**
Applications	Adhesives Coatings	Adhesives Coatings Electronics	Bonds Fibrous Materials	High-Purity Binder
pH	Mildly Basic	Mildly Basic	Mildly Basic	Mildly Acidic
Maximum Temperature (°C)	1650	1650	1760	1870
Density (g/cm^3)	1.40	1.20	1.42	1.40
Viscosity (cps)	1500	150	50	500
Cure Time (hours)	24	24	24	2
Cycling Instructions	16 hours 25 °C or 2 hours at 120 °C	16 hours 25 °C or 2 hours at 120 °C	2 hours 25 °C or 2 hours at 175 °C	2 hours 95 °C and 2 hours at 175 °C

SAMPLE PREPARATION

In order to create the TSPs, phosphor powder is mixed with ceramic binder. The compositions for paint samples used in this research are shown in Table III. Typically, the phosphor powder accounts for twenty percent of the paint mixture by volume. In some cases, ten percent (by volume) of a second powder such as magnesium oxide (MgO_2) is added displacing the binder. This addition was an attempt to increase the thermal conductivity to reduce thermal shock, increase the survivability of the coating, and is discussed in reference 13.

Table III. Tested paint sample compositions (fractions by volume).

Resbond Binder		Phosphor		Water	MgO_2
Product	Quantity	Compound	Quantity	Quantity	Quantity
791	35%	Y_2O_3:Eu	20%	35%	10%
	40%			40%	0%
	70%			0%	10%
	80%			0%	0%
792	70%	Y_2O_3:Eu	20%	0%	10%
	80%			0%	0%
793	80%	Y_2O_3:Eu	10%	0%	10%
			20%	0%	0%

The paint preparation procedure began by measuring the powdered solids in a graduated capped vial. Individual tubes were vibrated to insure proper volume measurement. Liquids were then added in predetermined amounts. After all the components were added to the cylinder, they were mixed by vigorously shaking the capped vial. The mixture was applied to the surface of a cleaned alumina substrate using a commercial airbrush. The TSPs were allowed to cure according to the manufacturer's specifications as shown in Table II. In the case of Resbond 791, water was added to thin the mixture to make it easier to apply using the airbrush. However the undiluted Resbond 791 paints were applied using a brush, since is difficult to spray and did not give the desired uniform coating.

EXPERIMENTAL METHOD

In order to determine the survivability for a TSP, samples were thermally cycled in a Thermoline 46200 high-temperature furnace. The furnace controls were set to raise the temperature to some predetermined value. The sample was then allowed to remain at this high temperature for one hour or more, followed by cooling to ambient room conditions. To quantify the fluorescence efficiency of phosphor suspended in the TSP, a Perkin Elmer LS-50B spectrophotometer was used to measure the emission spectrum of the sample after each thermal cycle. For each TSP mixture, two samples were made. The first sample was heated through the curing cycle and used as a control. The second was thermally cycled by heating to a set temperature and allowing it to cool back to room temperature. After the sample cooled, the emission spectrum was measured. Figure 2 shows the variation in the emission spectrum for a Resbond 793 and Y_2O_3:Eu-based TSP after thermal cycling at several different temperatures. It is quite obvious that fluorescence intensity for the 611 nm emission line decreases as a function of temperature. In fact, the 611 nm fluorescence emission is reduced to the background level after thermal cycling at 1600 °C. The TSP fluorescence spectrum shown in Figure 2 contained 80% Resbond 793 and 20% Y_2O_3:Eu by volume.

QUALITATIVE RESULTS

After thermal cycling, TSP samples were analyzed under a ultraviolet light to qualitatively estimate the intensity of the remaining 611 nm fluorescence from Y_2O_3:Eu. The TSPs were held at temperature for one hour. Qualitative results for the Resbond binder and

Y_2O_3:Eu TSP samples are shown in Table IV. Notice that all of the tested TSP samples showed some fluorescence after thermal cycling to 1300 °C. Fluorescence was also observed for all but one TSP sample at 1400 °C. Only three TSPs made with Resbond 792 and 793 showed significant fluorescence at 1500 °C. Finally, fluorescence was observed at 1600 °C for one TSP containing 20% Y_2O_3:Eu and 80% Resbond 793 by volume. These results are consistent with earlier TSP research completed during the second year of the NASA-Glenn program [ref. 7-12]. As a group, it appears that TSPs made with Zyp binders were better able to withstand slightly higher temperatures compared to the ones made using any of the three tested Cotronics Resbond formulations. However, selected Zyp and Cotronics based TSPs were both able to survive to 1500 °C.

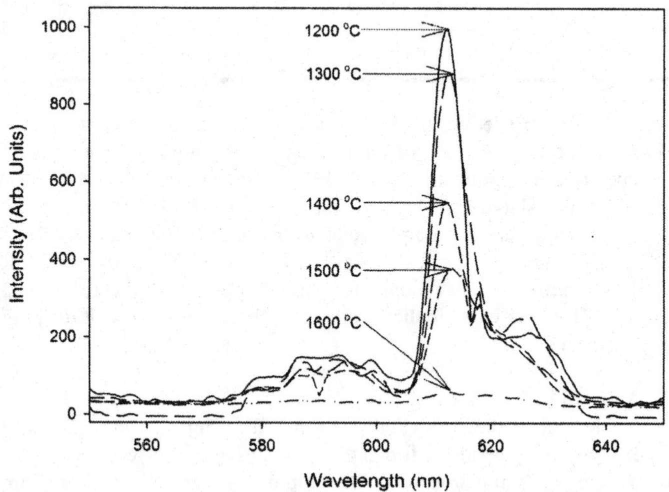

Figure 2. Resbond 793 and Y_2O_3:Eu TSP emission spectrum after thermal cycling.

Table IV. Qualitative results for the Resbond binder and Y_2O_3:Eu TSP samples.

Resbond Binder		Remaining Paint Components			Phosphor Emission At Given Cycling Temperature?				
Type	Amount	Y_2O_3:Eu	Water	MgO_2	1200 °C	1300 °C	1400 °C	1500 °C	1600 °C
791	35%	20%	35%	10%	Yes	Yes	Yes	No	No
	40%		40%	0%	Yes	Yes	Yes	No	No
	70%		0%	10%	Yes	Yes	Yes	No	No
	80%			0%	Yes	Yes	Yes	No	No
792	70%	20%	0%	10%	Yes	Yes	Yes	Yes	No
	80%			0%	Yes	Yes	Yes	Yes	No
793	70%	20%	0%	10%	Yes	Yes	No	No	No
	80%			0%	Yes	Yes	Yes	Yes	Yes

QUANTITATIVE RESULTS

A quantitative determination of fluorescence intensity as a function of cycling temperature is more complex. It was decided to use a ratio of 0.2 (20%) of the maximum emission intensity as the criteria to determine the viability of fluorescence for a given TSP sample. If the fluorescence emission is small, it will be difficult to measure the decay time and obtain a corresponding surface temperature. There will come a point in intensity where a phosphor system cannot be used to measure temperature. The decision ratio of 0.2 was completely arbitrary and was based on the observation that the apparent fluorescence measurement uncertainty was about ± 10% (intensity fraction of 0.1), which was two times the measured error for the 611 nm line for Y_2O_3:Eu.

For each tested TSP, the spectral fluorescence intensity was measured using a Perkin Elmer LS-50B spectrophotometer and an ultraviolet excitation source. In each case, the fluorescence was normalized to the Y_2O_3:Eu maximum emission line at 611 nm. Samples were held at temperature for one hour unless otherwise indicated. Quantitative results for each Resbond binder can be found in the sections that follow.

Resbond 791

The evaluation of the Resbond 791 binder was complicated by its large viscosity. This required modifying the application procedure by either applying the paint with a standard bristle brush, increasing the supply pressure to the airbrush, or by thinning the binder. Since the other evaluated Resbond binders had smaller viscosities, it was felt that applying the thick paint with a brush and thinning of binder would provide a more equitable comparison. As such, the Resbond 791 binder was thinned in a one-to-one ratio with water, which aided in the application of the paint. Quantitative results for the Resbond 791 binder TSP are shown in Figure 3.

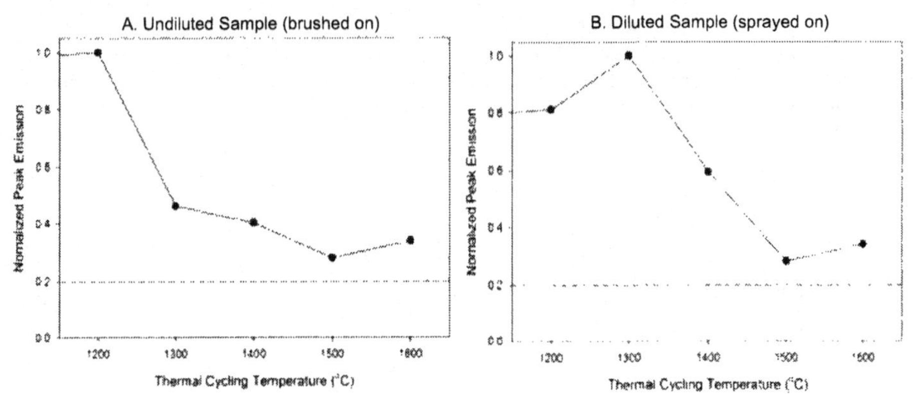

Figure 3. Normalized emission for brushed on and sprayed on Resbond 791 TSPs as a function of cycling temperature.

Tests were performed using brushed on paints. These coatings were thicker than the airbrushed coatings, but appeared to have the same maximum cycling temperature. The plot on the left side of Figure 3 (labeled A) shows the normalized fluorescence emission (611 nm) for the brushed on Resbond 791 paint versus cycling temperature. The brushed on TSP maintained a normalized peak emission greater than 0.2 to cycling temperatures up to 1600 °C. The emission intensity decreased consistently from 1300 to 1500 °C.

The plot on the right side of Figure 3 (labeled B) shows the normalized fluorescence emission (611 nm) for the airbrushed (diluted) Resbond 791 paint versus cycling temperature. The normalized emission spectra for the diluted sample was similar to that measured for the brushed on TSP. It is interesting to note that the intensity of the brushed on samples decreases slightly faster than the diluted TSPs, which could be partially caused by differences in application method. The undiluted samples were brushed on to the substrate surface that typically does not provide as even a coating. The uneven thickness could cause uneven heating and enhance flaking of the TSP.

Resbond 792

For all tested cases, the Resbond 792 TSPs falls below the 0.2 intensity criteria after the first thermal cycle (1100 °C). Visual inspection of the samples in ultraviolet light indicates that there was a significant drop in emission intensity after 1100 °C. These inspections also showed discernable fluorescence emission to 1300 °C and limited fluorescence out to 1500 °C. The dramatic drop in fluorescence intensity would tend to indicate that the Resbond 792 is not suitable for high-temperature TSP applications.

Resbond 793

Quantitative results for the Resbond 793 binder TSP are shown in Figure 4. The Resbond 793 TSP maintained a normalized peak emission greater than 0.2 to cycling temperatures up to 1500 °C. The corresponding 611 nm normalized peak emission decreased uniformly from 1200 to 1600 °C. Visual inspection of the samples indicates that the coating remains relatively bright to 1400 °C. At 1500 °C, the coating is still bright but is starting to flake away from the surface. By 1600 °C, almost all of the coating has flaked off of the surface, but the remaining portion was still discernable. Resbond 793 has a low viscosity, which makes it easy to spray, and has the highest temperature survivability of any of the tested Cotronics binders. It appears to be the best of the tested materials for use in a TSP.

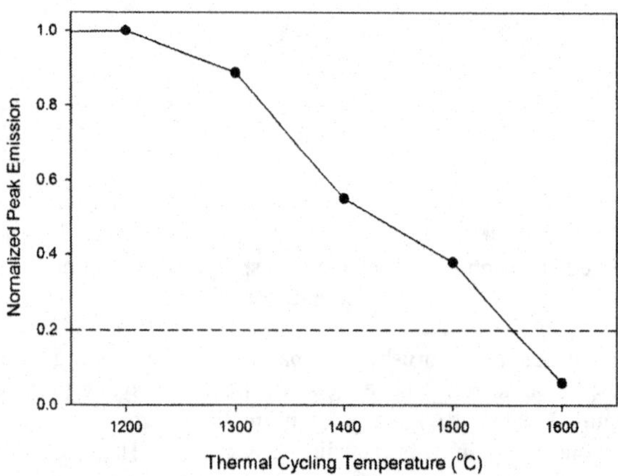

Figure 4. Normalized emission for Resbond 793 TSP as a function of cycling temperature.

Temperature Exposure Time Dependence

The exposure time was originally selected because the anticipated test length was only a few seconds. It was felt that an hour of temperature exposure would well exceed the lifetime of any experimental test series. This exposure time was selected to make this work consistent with earlier research [ref. 7-12]. To date, no research had been completed comparing the effects of exposure time at temperature to the magnitude of fluorescence emission. It was decided to complete a series of tests with a single binder to determine the relationship between temperature exposure time and normalized emission intensity for a Y_2O_3:Eu and Resbond 793 TSP.

To complete these tests, four new samples were prepared and cured as described in earlier sections. The first sample was held in reserve as the post cure control. The remaining samples were heated to 1400 °C for 1, 2, and 4 hours. The 1400 °C temperature was selected because nearly half of the room temperature fluorescence was emitted for this TSP in the previous test series. A separate sample was used in each duration test. The resulting emission spectra (ultraviolet excitation) for Y_2O_3:Eu in a Resbond 793 binder with varied thermal cycling duration are shown in Figure 5. After a one hour exposure, there is a noticeable decrease in fluorescence intensity. In fact, the decrease in intensity in the first hour of the test is surprising since the previous results showed a 40% decrease, while this test had a nearly 70% decrease from the control. After two hours, there is further decrease in intensity to greater than 80% of the control, which indicates that the coating is marginal after two hours of exposure to 1400 °C. After four hours, almost all the fluorescence is gone.

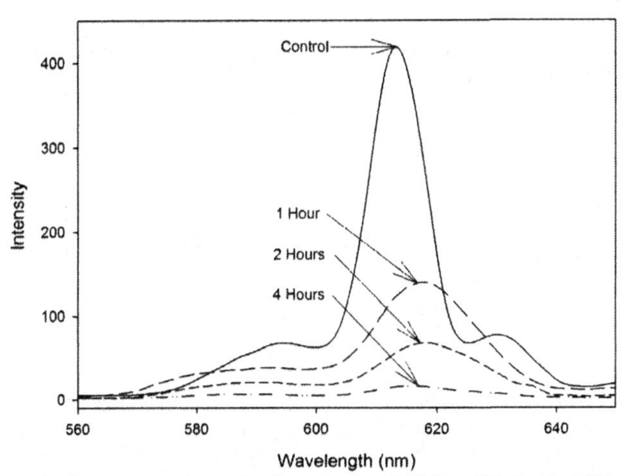

Figure 5. Changes in fluorescence intensity for a Resbond 793 and Y_2O_3:Eu TSP as a function of thermal exposure time.

CONCLUSIONS

Phosphors are fine powders that are doped with trace elements that give off visible light when suitably excited. Some phosphors can survive and function at high temperatures such as those present during combustion. This research provided basic results as to which Resbond-based TSPs can emit fluorescence and remain mechanically viable to high temperatures. All of the tested Resbond-based Y_2O_3:Eu TSP samples showed some fluorescence after thermal cycling to 1300 °C. Fluorescence was also observed for all but one TSP sample at 1400 °C. Only three TSPs made with Resbond 792 and 793 showed significant fluorescence at 1500 °C. Fluorescence was observed at 1600 °C for one TSP containing Resbond 793. These results are consistent with similar survivability measurements completed with other TSPs. In addition, Resbond 793 appears to have the best TSP characteristics of the Cotronics binders, and it is one of the few binders that survives above 1500 °C. The temperature exposure time was found to be an important consideration for designing a TSP.

After two hours of exposure at 1400 °C, the fluorescence intensity decreases by 80%, which indicates the coating is of marginal use as a TSP. It appears that TSPs made with Zyp binders were better able to withstand slightly higher temperatures compared to the ones made using any of the three tested Cotronics Resbond formulations. However, selected Zyp and Cotronics based TSPs were both able to survive to 1500 °C. Additional research is needed to further quantify these results.

ACKNOWLEDGEMENTS

The authors are grateful for the financial support provided by NASA Glenn Research Center through an interagency agreement with the U.S. Department of Energy. This paper was prepared by the Oak Ridge National Laboratory, Oak Ridge, Tennessee, managed by UT-BATTELLE, LLC, for the U.S. Department of Energy under contract DE-AC05-00OR22725. Accordingly, the U.S. government retains a nonexclusive, royalty-free license to publish or reproduce these documents, or to allow others to do so, for U.S. government purposes. The Louisiana Education Quality Support Fund (LEQSF) using grant using LEQSF (2000-03)-RD-A-39 provided additional support for this research.

REFERENCES

1. S.W. Allison, M.R. Cates, S.M. Goedeke, W.A. Hollerman, F.N. Womack, and G.T. Gillies, "Remote Thermometry With Thermographic Phosphors: Instrumentation and Applications", Chapter 4, Handbook of Luminescence, Display Materials, and Devices, Volume 2: Inorganic Display Materials, Edited by H.S. Nalwa and L.S. Rohwer, American Scientific Publishers, 187-250 (2003).

2. S.W. Allison and G.T. Gillies, "Remote Thermometry with Thermographic Phosphors Instrumentation and Applications", Rev. Sci. Instrum, 68 (7), 2615-2650, 1997.

3. M.R. Cates, S.W. Allison, L.A. Franks, H.M. Borella, B.R. Marshall, and B.W. Noel, "Laser-Induced Fluorescence of Europium-Doped Yttrium Oxide for Remote High-Temperature Thermometry", Proc. Laser Inst. Am. 49-51, 142, 1985.

4. S.W. Allison, L.A. Boatner, G.T. Gillies, "Characterization of High-Temperature Thermographic Phosphors: Spectral Properties of $LuPO_4$:Dy (1%)Eu(2%)," Appl. Opt. 34, 5624, 1995.

5. O.A. Lopez, J. McKittrick, L.E. Shea, "Fluorescence Properties of Polycrystalline Tm^{+++}-Activated $Y_3Al_5O_{12}$ and Tm^{+++}- Li^+ Co-activated $Y_3Al_5O_{12}$. in the Visible and Near IR Ranges," Journal of Luminescence 71, 1-11, 1997.

6. S.W. Allison, D.L. Beshears, T. Bencic, W.A. Hollerman, and P. Boudreaux, "Development of Temperature-Sensitive Paints for High Temperature Aeropropulsion Applications", Proceedings of the American Institute of Aeronautics and Astronautics Propulsion Conference, AIAA-2001-3528, 2001.

7. S.W. Allison, D.L. Beshears, T. Gadfort, T. Bencic, J. Eldridge, W.A. Hollerman, and P. Boudreaux, "High Temperature Surface Measurements Using Lifetime Imaging of Thermographic Phosphors: Bonding Tests", 19th International Congress on Instrumentation in Aerospace Simulation Facilities, August 27-30, 2001.

8. W.A. Hollerman, S.W. Allison, D.L. Beshears, R.F. Guidry, T.J. Bencic, C.R. Mercer, J.I. Eldridge, M.R. Cates, P. Boudreaux, and S.M. Goedeke, "Development of Fluorescent Coatings for High Temperature Aerospace Applications", Proceedings of the 2002 Core Technologies for Space Systems Conference Proceedings, Colorado Springs, CO., November 19-21, 2002.

9. W.A. Hollerman, S.W. Allison, D.L. Beshears, R.F. Guidry, T.J. Bencic, C.R. Mercer, J.I. Eldridge, M.R. Cates, P. Boudreaux, and S.M. Goedeke, "Development of Inorganic Fluorescent Coatings for High Temperature Aerospace Applications", Proceedings of the 49th International Instrumentation Symposium, Orlando, Florida, May 5-9, 2003.

10. W.A. Hollerman, S.W. Allison, S.M. Goedeke, P. Boudreaux, R. Guidry, and E. Gates, "Comparison of Fluorescence Properties for Single Crystal and Polycrystalline YAG:Ce", IEEE Transactions on Nuclear Science, 50 (4), 754-757, 2003.

11. S.W. Allison, S.M. Goedeke, D.L. Beshears, M.R. Cates, W.A. Hollerman, F.N. Womack, N.P. Bergeron, T.J. Bencic, C.R. Mercer, and J.I. Eldridge, "Advances In High Temperature Phosphor Thermometry For Aerospace Applications", Proceedings of the 39th AIAA, ASME, SAE, ASEE Joint Propulsion Conference, Huntsville, AL, July 20-23, 2003.

12. W.A. Hollerman, R.F. Guidry, F.N. Womack, N.P. Bergeron, S.W. Allison, S.M. Goedeke, D.L. Beshears, M.R. Cates, T.J. Bencic, C.R. Mercer, and J.I. Eldridge, "Use of Phosphor Coatings For High Temperature Aerospace Applications", Proceedings of the 39th AIAA, ASME, SAE, ASEE Joint Propulsion Conference, Huntsville, AL, July 20-23, 2003.

13. N.P. Bergeron, W.A. Hollerman, S.M. Goedeke, S.W. Allison, M.R. Cates, T.J. Bencic, C.R. Mercer, and J.I. Eldridge, "Effect of Adding MgO_2 to a Selection of Resbond® Thermally Sensitive Paints", Proceedings of the 50th Annual International Instrumentation Symposium, San Antonio, TX, May 9-13, 2004.

Author Index

Keyword Index

9 781574 981841